MW01035495

Texts and Monographs in Computer Science

The AKM Series in Theoretical Computer Science

A Subseries of Texts and Monographs in Computer Science

A Basis for Theoretical Computer Science
by M. A. Arbib, A. J. Kfoury, and R. N. Moll

A Programming Approach to Computability
by A. J. Kfoury, R. N. Moll, and M. A. Arbib

An Introduction to Formal Language Theory
by R. N. Moll, M. A. Arbib, and A. J. Kfoury

A Programming Approach to Computability

A. J. Kfoury
Robert N. Moll
Michael A. Arbib

Springer-Verlag
New York Heidelberg Berlin

A. J. Kfoury
Department of Mathematics
Boston University
Boston, MA 02215
U.S.A.

Robert N. Moll
Department of Computer and
Information Science
University of Massachusetts
Amherst, MA 01003
U.S.A.

Michael A. Arbib
Department of Computer and
Information Science
University of Massachusetts
Amherst, MA 01003
U.S.A.

Series Editor

David Gries
Department of Computer Science
Cornell University
Upson Hall
Ithaca, NY 14853
U.S.A.

On the front cover is a likeness (based on a postcard kindly supplied by Professor S. C. Kleene) of the ninth century mathematician Abu Jafar Muhammad Ibn Musa Al-Khowarizmi. The word "algorithm" is derived from the last part of his name.

With 36 Figures

Library of Congress Cataloging in Publication Data
Kfoury, A. J.
A programming approach to computability.
(The AKM series in theoretical computer science)
(Texts and monographs in computer science)
"February 1982."
Bibliography: p.
Includes index.
1. Computable functions. 2. Programming
(Electronic computers) I. Moll, Robert N.
II. Arbib, Michael A. III. Title. IV. Series.
V. Series: Texts and monographs in computer science.
QA9.59.K46 1982 519.4 82-10488

Printed in the United States of America.

9 8 7 6 5 4 3 2 1

ISBN 0-387-90743-2 Springer-Verlag New York Heidelberg Berlin
ISBN 3-540-90743-2 Springer-Verlag Berlin Heidelberg New York

Preface

Computability theory is at the heart of theoretical computer science. Yet, ironically, many of its basic results were discovered by mathematical logicians prior to the development of the first stored-program computer. As a result, many texts on computability theory strike today's computer science students as far removed from their concerns. To remedy this, we base our approach to computability on the language of **while**-programs, a lean subset of PASCAL, and postpone consideration of such classic models as Turing machines, string-rewriting systems, and μ-recursive functions till the final chapter. Moreover, we balance the presentation of unsolvability results such as the unsolvability of the Halting Problem with a presentation of the positive results of modern programming methodology, including the use of proof rules, and the denotational semantics of programs.

Computer science seeks to provide a scientific basis for the study of information processing, the solution of problems by algorithms, and the design and programming of computers. The last 40 years have seen increasing sophistication in the science, in the microelectronics which has made machines of staggering complexity economically feasible, in the advances in programming methodology which allow immense programs to be designed with increasing speed and reduced error, and in the development of mathematical techniques to allow the rigorous specification of program, process, and machine. The present volume is one of a series, the AKM Series in Theoretical Computer Science, designed to make key mathematical developments in computer science readily accessible to undergraduate and beginning graduate students. The book is essentially self-contained—what little background is required may be found in the AKM volume *A Basis for Theoretical Computer Science*.

The book is organized as follows. Chapter 1 provides a preview of the techniques to be used throughout the book in determining whether or not a function is computable by some algorithm.

Chapters 2 and 3 establish our framework for the study of computability, using the lean PASCAL subset we call "**while**-programs". In these two chapters we take a close look at the intuitive idea of an algorithm as formulated in this simple programming language. We also introduce the idea of a universal program or interpreter which can simulate any program P when given P's data together with an encoding of P as an additional input.

In Chapter 4 we develop basic techniques of computability theory, which we then use to examine the computational status of various natural problems.

Chapter 5 provides an introduction to program correctness and to alternative theories of semantics of programming languages.

In Chapter 6 we consider the role of self-reference in computability theory as embodied in a result known as the Recursion Theorem. We also study conditions under which the theory can be made model independent.

In Chapters 7 and 8 our emphasis changes from functions to sets. Recursive sets, whose characteristic functions are computable, are compared with recursively enumerable sets whose elements are listable by algorithms. As we shall see, these two concepts do not coincide.

Finally, in Chapter 9 we examine alternative formulations of computability theory, and we show how one of these formulations (the Turing machine model) can be used to embed the results of computability theory in a large variety of other symbol-processing systems, including formal languages.

More detailed outlines may be found at the beginning of each chapter.

In keeping with the style of the AKM series, we have resisted the temptation to include more material than could possibly be covered in a one-semester course. Nevertheless, in moderating the pace for a single term's work, the instructor may find it appropriate to omit material from one or more of Sections 5.3, 6.2, 7.3, 8.2, 9.2, and 9.3 (the starred sections in the table of contents).

The book grew out of our teaching classes at the University of Massachusetts at Amherst and at Boston University over several years. We owe special thanks to Ernest Manes for letting us use material from the forthcoming text *Formal Semantics of Programming Languages* by M. A. Arbib and E. Manes in the preparation of Section 5.1. We thank our students for helping us integrate computability theory into the "world view" of modern computer science, and we thank Susan Parker for all her efforts in typing the manuscript.

February 1982

A. J. Kfoury
R. N. Moll
M. A. Arbib

Contents

*These sections may be omitted with no loss of continuity.

CHAPTER 1

Introduction

Computability theory looks at the functions computed by algorithms, such as computer programs. Since a program may not halt for a given input, the functions so computed will be partial functions. In Section 1.1 we discuss partial functions and their computability by different families of algorithms. In Section 1.2 we give an example of an unsolvability or noncomputability result—showing that no program can compute for an arbitrary number n whether or not the PASCAL program with ASCII code n will halt when also given n as input data. Section 1.3 shows that the method of proof used to show the noncomputability of this halting problem is akin to Cantor's diagonal argument that the real numbers are uncountable. This chapter thus gives the reader a taste for a number of techniques to be used again and again throughout this volume.

1.1 Partial Functions and Algorithms

A function is computable if it is determined by some algorithm, that is, by some well-specified program or set of rules. In this section we introduce the notion of a partial function—the proper functional concept for computability theory—and then look closely at the relationship between algorithms and functions.

Functions

We assume that the reader already knows what the word "function" means. (The necessary background for this section is provided in the companion volume, *A Basis for Theoretical Computer Science*, M. A. Arbib, A. J. Kfoury, and R. N. Moll, Springer-Verlag, New York, 1981, Chapter 1.) Recall that in conventional mathematics, a function $f: A \to B$ is some rule according to which each element x of A (the *domain* of arguments) is associated with an element $f(x)$ in another set B (the *codomain* of values). This rule may be a precise algorithm, or merely a hypothetical rule of correspondence. Most common functions of everyday arithmetic, e.g., the multiplication function $g(x, y) = x * y$, have familiar algorithmic descriptions, and are routinely classified as computable functions. However, the status of a function specified by a hypothetical rule of correspondence may be less clear. As an example, consider the function f given below mapping domain $\mathbf{N}$, the set of non-negative integers, to codomain $\{0, 1\}$.

$$f(x) = \begin{cases} 1, \text{ if Fermat's last theorem}^1 \text{ is true;} \\ 0, \text{ if Fermat's last theorem is false.} \end{cases}$$

Now mathematicians have not yet established the validity of Fermat's last theorem. So it is not known if $f(x)$ is the constant function identically equal to zero, or the constant function identically equal to 1. In either case, however, f can be calculated by a trivial algorithm—either "output 0 on all arguments," or "output 1 on all arguments." Hence, f is considered to be computable even though the algorithm for calculating its value has not yet been identified.

As every programmer knows, a poorly designed but runnable algorithm will often fail to terminate for some input values, and our theory must take this phenomenon into account. Consider, for example, the function computable by the algorithm shown in the flow chart of Figure 1, and which is expressed as follows in the computer language PASCAL[2]:

```
begin
    while X mod 2 = 0 do
    X := X + 2
end
```

[1] Fermat's last theorem is the as yet unproved assertion that for any $n > 2$, the equation $x^n + y^n = z^n$ has no integer solutions.

[2] Our convention for writing variables throughout the book is this: We use lower case letters, mostly from the end of the Roman alphabet, e.g., x_1, x_2, y, y_{12}, z, etc.—except when these variables appear in the text of programs, in which case we use the corresponding upper case letters, e.g., $X1$, $X2$, Y, $Y12$, Z, etc.

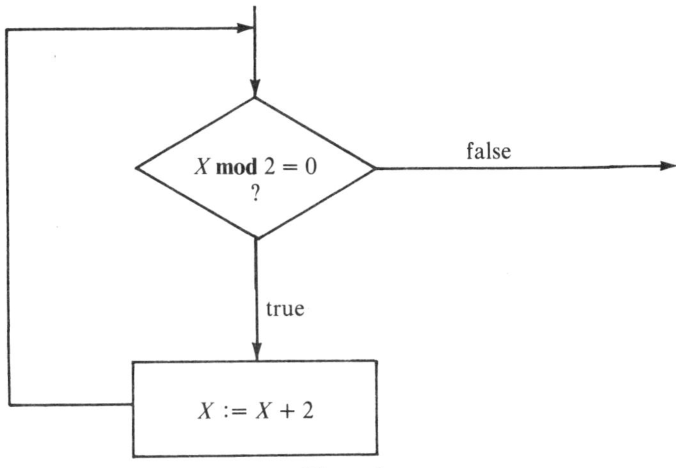

Figure 1

Let us assume that this algorithm determines a function θ from $\mathbf{N}$ to $\mathbf{N}$. If we think of X as the input and output variable, then the function θ may be described as follows:

$$\theta(x) = \begin{cases} x, & \text{if } x \text{ is odd;} \\ \text{undefined}, & \text{otherwise.} \end{cases}$$

That is, $\theta(3) = 3$, but $\theta(4)$ has no value, since on this input the program enters an infinite loop and runs forever. This phenomenon—undefined or *divergent* computations—is a fundamental feature of computability theory. Accordingly we must adjust our notation of functional relationships in order to make this aspect of our computability theory mathematically tractable.

1 Definition. A *partial function* is a function that may or may not be defined on all its arguments. Specifically, whereas an "ordinary" function $g: A \to B$ assigns a value $g(x)$ in B to each x in A, a partial function $\theta: A \to B$ assigns a value $\theta(x)$ only to x's in some subset DOM(θ) of A. DOM(θ) is called the *domain of definition* of θ. If $x \notin \text{DOM}(\theta)$ we say that θ is undefined or unspecified at that value. Note that we shall refer to A as the *domain* of $\theta: A \to B$, and to B as its *codomain*. The set $\{\theta(x) \mid x \in \text{DOM}(\theta)\}$ is called the *range* of θ, and is denoted by RAN(θ).

Our example function above, mapping $\mathbf{N}$ to $\mathbf{N}$, is a partial function with domain DOM(θ) = {non-negative odd integers}; θ is undefined on all even arguments. In this example, the range of θ equals its domain of definition.

Since the empty set $\emptyset$ is a subset of any set, it is a valid partial function domain. Hence the function $\perp$ defined by

$$\perp(x) = \text{undefined}, \qquad \text{for all} \quad x \in \mathbf{N}$$

is a valid partial function, which we call the *empty function*. It is also a computable function, and the algorithm of Figure 2—below, expressed in PASCAL—computes it:

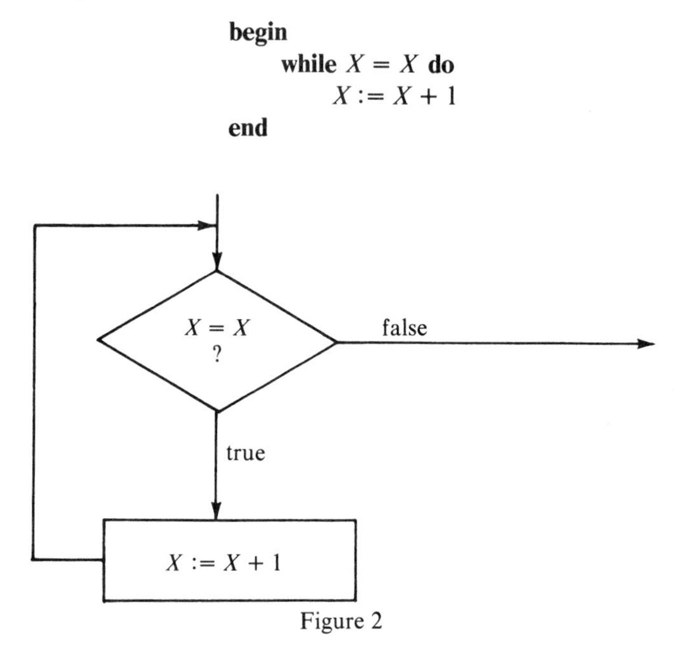

Figure 2

In later chapters we shall see that the empty function plays a very useful role in the development of computability theory.

2 Definition. A *total function* $f: A \to B$ is a partial function with domain of definition $\mathrm{DOM}(f) = A$.

A total function is thus a bona-fide function in the conventional mathematical sense. Like the empty function, total functions are of great importance for computability theory.

Throughout this book when we talk about functions we will in general mean partial functions, and we will usually denote partial functions by lower case Greek letters. Often we will discover or observe that a particular partial function is indeed total, i.e., defined for all its arguments. We will often record this fact by naming the function with a lower case Roman letter.

Let us now introduce set notation for functions. We first define what relations are.

3 Definition. For an integer $k > 0$, a *k-ary relation* R, between sets A_1, $A_2, \ldots, A_k$ is simply a subset of the Cartesian product $A_1 \times A_2 \times \cdots \times A_k$ of the k sets.

If $\mathbf{N}$ denotes the set of all non-negative integers, an example of a 2-ary or binary relation on $\mathbf{N}$ (i.e., $A_1 = A_2 = \mathbf{N}$) is:

$$\{(m,n) \mid m \in \mathbf{N},\, n \in \mathbf{N}, \text{ and } m < n\},$$

which denotes a set of ordered 2-tuples (or pairs) of the form (m,n) satisfying the conditions listed on the right-hand side of the vertical bar. A partial listing of this relation is thus

$$(0,1), (0,2), (0,3), \ldots, (1,2), (1,3), \ldots, (2,3), (2,4), \ldots, (3,4), \ldots .$$

An alternative definition of a partial function $\theta : A \to B$ is to say θ is a relation between A and B such that, for every $x \in A$, there exists at most one $y \in B$ such that $(x, y) \in \theta$.

Several operations on functions will be used in the course of this book, and most will be introduced in due time. The first is so basic that we define it here.

4 Definition. *Composition of functions* $\theta : A \to B$ and $\zeta : B \to C$ (with the codomain of the first matching the domain of the second) is the operation which yields the *composite* $\zeta \circ \theta : A \to C$ defined to be

$$\{(x,z) \mid \text{there is a } y \in B \text{ such that } (x, y) \in \theta \text{ and } (y,z) \in \zeta\}.$$

We may also define $\zeta \circ \theta$ simply by writing:

$$\zeta \circ \theta(x) = \zeta(\theta(x)),$$

which means: first evaluate $\theta(x)$, and if this is undefined then $\zeta(\theta(x))$ is undefined; otherwise evaluate ζ for the argument $\theta(x)$ and the result, whether defined or not, is the value of $\zeta \circ \theta(x)$. Thus $\mathrm{DOM}(\zeta \circ \theta) = \{x \mid x \in \mathrm{DOM}(\theta) \text{ and } \theta(x) \in \mathrm{DOM}(\zeta)\}$.

We conclude this subsection by placing a restriction on all the functions to be considered in this book: Unless stated otherwise all functions will be of the form

$$\theta : \mathbf{N}^k \to \mathbf{N}^l.$$

That is, given k numerical values for input, the function θ will (if defined) return l values for output. Usually, l will be 1. (This restriction does not limit the scope of the theory we develop. After all, any data can be stored in a computer in a binary code, and each binary pattern may be interpreted as a natural number.)

Algorithms

Computer scientists usually associate the word "algorithm" with computer programs, and we relied on this association in our introduction to partial functions earlier in the section. But algorithmic ideas are also present at the most elementary levels of arithmetic: "To know how to add", "to know how to multiply", "to know how to divide", all mean to know certain algorithms. Although they are rarely mentioned by name—a notable exception is when we refer to Euclid's Algorithm for finding the greatest common divisor of two integers—algorithms are in fact found throughout mathematics. And algorithms are also present in the most practical situations of everyday life; e.g., instructions for any do-it-yourself equipment, a cooking recipe, a medical prescription, etc.

If we do not often use the word "algorithm", it is perhaps because we have a large variety of equivalent terms: a "sequence of instructions", a "recipe", a "prescription", a "method", a "plan", etc. In more technical settings, we meet other equivalent terms: a "program", an "effective procedure", etc. This multiplicity of words and expressions reflects the fact that the concept of an algorithm, in contrast to the concept of a function, does not seem to have a simple, unified definition.

Let's agree on some conditions for the existence of an algorithm before attempting to give a definition of this concept. An algorithm works on some information, the input data, to produce some other information, the output data. And for every algorithm, it is only a restricted set of possible input data to which the application of the algorithm at hand becomes meaningful. For example, in the case of Euclid's Algorithm that finds the g.c.d. (greatest common divisor) of two numbers, the input data must be a pair of positive integers. Now, an algorithm may or may not work properly on input data that are not of the appropriate kind, in the sense that it may or may not produce the corresponding output data. For example, Euclid's Algorithm never terminates if one or both of its input data is a negative integer.

Moreover, the input to an algorithm must have a finite description. And because an algorithm must start from a finite object, and must run for a finite amount of time if an output is to be obtained, it can only construct another finite object. Thus another condition for the existence of an algorithm is that both its input and output have finite descriptions.

We also insist that an algorithm must lead to the same outcome every time it is started on the same input. In this way an algorithm establishes a functional relationship between its set of possible input data and its set of possible output data. The (partial) function so defined is a computable function.

Now let us turn to the actual specification of algorithms. Several independent attempts have been made to provide a most general definition of an algorithm; each specification yields a corresponding class of computable functions. Two of the best known attempts are:

1. The Turing-machine description of algorithms.
2. The Kleene characterization of computable functions.

Another approach, to be adopted in this book, is what we may call:

3. The program description of algorithms.

By "program description" we mean a high-level programming language with a strong resemblance to several of the programming languages encountered in the practice of computer science. The full definiton of our high-level programming language will be given in Chapter 2; as for the two other approaches mentioned above, they will be discussed in Chapter 9.

If a specification of algorithms is indeed most general then it must be *complete* in the following sense: Every conceivable algorithm, independent of its form of expression, must be equivalent to some algorithm written in the language of the complete system. Now, what do we mean by the "equivalence" of two algorithms? For the moment, let us simply say that two algorithms are *equivalent* in the case that they compute the same function; later, this preliminary definition of equivalence will be made more precise. What is important here is that no one has ever come up with an algorithm that cannot be replaced by an equivalent algorithm in any of the three aforementioned specification systems. In particular, the three systems can be proven to define exactly the same computable functions. More strongly we can say, in the terminology of computer science, that for each pair of formalisms there is a *compiler* which can translate each algorithm in one formalism to an equivalent algorithm in the other formalism. (More on this in Chapter 9.)

Although we do not have a proof for the completeness of the three forementioned specifications of algorithms, we do know they are as complete as any specification yet invented. We can abstract some of their common features, which we can then agree must be present in *any* system of algorithmic specification. These features are in fact quite simple, and we may briefly mention them here (in terms which will become quite precise and rigorous as the book progresses):

An algorithmic process, which is the process of applying an algorithm to appropriate input data, can be decomposed into successive elementary steps. Each elementary step consists of replacing one state of the process by another. This replacement of one state by the next is carried out in a deterministic, mechanical manner, according to elementary rules of transformation. The outcome of an algorithmic process may be either (i) replacing one state by the next without ever stopping, or (ii) reaching a state for which there is no next state according to the transformation rules. Only in the case (ii) can we obtain output data from the algorithmic process.

Let us point out, without further elaboration, that the word "elementary" in the preceding paragraph is highly imprecise. What we mean by "elementary steps" and "elementary rules" is that they will not be decomposed into simpler steps and simpler rules; but we still have to agree on the nature of what is to be indecomposable into simpler components.

Church's thesis

Each available specification system for algorithms—any of the three mentioned earlier or any of several others[3]—only allows very simple "elementary steps" and very simple "elementary rules" of transformation. Algorithms in all these systems are indeed very simply defined, as will be plainly illustrated in the case of the algorithms written in our basic programming language.

This raises the question of whether the formalism of any one of these specification systems is adequate to capture fully the notion of an algorithm. Is there some clever way to define a more powerful formalism so that we can somehow obtain a new kind of algorithm? That is, are there algorithms that compute functions beyond the computing power of any of the previously specified algorithms?

In fact, every formalism for describing algorithms so far invented can only specify partial functions shown to be computable by Turing, Kleene, and program-specification approaches. This, together with the fact that these three specifications of algorithms are equivalent, leads to the following hypothesis named in honor of the American logician Alonzo Church.

5 Church's Thesis. *The Turing, Kleene, and program specifications of algorithms are each complete.*

This means that any algorithmically defined function can in fact be computed by an algorithm in every one of these specifications. As we develop the theory of computability, we shall gradually accumulate the evidence for this thesis. We shall also see that it is a very useful and powerful tool.

Exercises for Section 1.1

1. A partial function ψ is an extension of a partial function τ, in symbols $\tau \leqslant \psi$, if τ viewed as a set of ordered pairs is a subset of ψ. Construct a proper infinite ascending chain of partial functions $\psi : \mathbf{N} \to \mathbf{N}$. That is, construct a sequence

$$\psi_1 < \psi_2 < \psi_3 < \cdots < \psi_n < \cdots,$$

i.e., each ψ_n is a proper subset of ψ_{n+1} for every $n \geqslant 1$.

2. Let $\mathbf{Q}$ be the set of rational numbers, and let $\theta : \mathbf{Q} \times \mathbf{Q} \to \mathbf{Q}$ be defined by $\theta(a,b) = a/b$, the result of dividing a by b.

   (a) Is θ total? In any case, what is DOM(θ)?
   (b) What is RAN(θ)?

[3] Among these, there are three that have special historical importance and led to distinctive approaches to computability theory: Markov algorithms [Markov (1954)], Post production systems [Post (1943)], and Church's lambda-calculus [Church (1932)]. For more accessible presentations of some or all of these approaches, consult Arbib (1969), Brainerd and Landweber (1974), or Minsky (1967). For a lively historical account by one of the pioneers in the field, see Kleene (1981).

3. Define $f: \mathbf{N} \to \mathbf{N}$ by

$$f(n) = 1 + 3 + 5 + \cdots + (2n - 1).$$

Give a simple description of RAN(f). Prove your result by induction.

1.2 An Invitation to Computability Theory

In this section we present an informal version of the principal elementary result of computability theory, the *recursive unsolvability of the Halting Problem*. This result addresses the following question: Can there be an algorithm for deciding if an arbitrary computer program halts on an arbitrary input? The answer to this question is "no", not if we require a uniform, general method—a single computer program—as our decision agent. We shall prove a special case of this theorem; the result in full generality will follow in Section 4.3.

The setting for our result will be a slightly modified version of the computer language PASCAL. The informal and, to some extent, imprecise approach of this section will give way to a more careful treatment in Chapter 2. There we introduce a simpler and more tractable language, a PASCAL subset we call the language of **while**-programs. Our PASCAL variant here follows the description of the language given by Jensen and Wirth (1974) with the following exceptions. We use one additional standard type, "non-negative integer." Every main program must contain at least one variable of this type. Unless we state otherwise, we shall assume in the text that all variables are of type non-negative integer. All main programs have exactly one read statement, and exactly one write statement, and both statements have single arguments of type non-negative integer. Programs receive input through the read statement which must follow the first **begin** of the program text, and transmit output, if any, through the write statement which must immediately precede the last **end** of the program text. Thus every legal program in this section computes a partial function $\varphi: \mathbf{N} \to \mathbf{N}$ via the read and write statements. The value of φ on some input n is the value printed by the write statement, if the write statement is executed. If the write statement is not executed, or if the argument of the write statement has not been assigned a value, then $\varphi(n)$ is undefined.

We assume that a standard PASCAL compiler can be modified so that these changes to PASCAL become conditions on program legality. We assume that there is no fixed upper bound on the amount of memory available for any particular computation, and that there is no upper bound on program run time.

The first order of business in establishing our main result is the construction of a systematic list of all PASCAL programs. We accomplish this in the following way. Let

$$P_0, P_1, P_2, \ldots, P_n, \ldots$$

be our listing, where program P_n is identified as follows: Given n, convert n to binary notation, and divide this binary sequence into eight-bit ASCII[4] blocks. If such a division is not possible (as with P_4, i.e., program "100"), then P_n will be a standard program for the empty function

```
begin
    read(X);
    while X = X do
    X := X + 1;
    write(X)
end
```

Otherwise, the ASCII interpretation of n is fed to our PASCAL compiler. If P_n is not acceptable to the compiler, then P_n is once again identified with the program for the empty function given above. Otherwise, P_n is the PASCAL program denoted in ASCII by n.

Notice that every legal PASCAL program appears somewhere in the list. Indeed, if P is a PASCAL program, and if it is stored in core in ASCII, then the appropriate sequence of ASCII character representations interpreted as a binary number determines P's *index*—P's position in our master program list.

Moreover, there is a clear, systematic relationship between a program and its index: Given any n, there is an algorithm—"write n in binary and treat this representation of n as a sequence of ASCII codes"—for determining the PASCAL program P_n.[5] And given any legal program P, there is a systematic way to determine P's index, namely, interpret P's ASCII image as a binary number.

Does an arbitrary program P_n halt when supplied with its own index as input? We now show that the computational status of this question—a special case of the Halting Problem—is nonalgorithmic. That is, we shall show that the total function $f: \mathbf{N} \to \mathbf{N}$ defined by

$$f(x) = \begin{cases} 1, & \text{if } P_x \text{ halts on input } x; \\ 0, & \text{if } P_x \text{ fails to halt on input } x; \end{cases}$$

[4] ASCII stands for American Standards Code for Information Exchange. The ASCII Code for the letter "A" is 01000001; the Code for "B" is 01000010, etc.

[5] Although the map $n \mapsto P_n$ is algorithmic, we have not called it "computable." The formal definition of a "computable function" in Chapter 2 will apply to any mapping from $\mathbf{N}^k$ to $\mathbf{N}^l$, regardless of the notational system we use to denote natural numbers. Put otherwise, computability theory is the theory of algorithmic functions on the natural numbers "up to notational equivalence". The map $n \mapsto P_n$ is a translation from one notational system to another (PASCAL programs as strings of symbols can be used for counting), and is not defined as a function from $\mathbf{N}$ to $\mathbf{N}$—and this is the reason why we do not formally identify it as a computable function.

is not computable by any PASCAL program, or for that matter, by any program in any programming language. We emphasize that the function f given above is a bona-fide total function mapping $\mathbf{N}$ to $\mathbf{N}$. Its value for any input is either a "0" or a "1". f's status as a function is not in dispute. What we are claiming is that no algorithm exists for determining f's values on all inputs.

We prove our result by contradiction. Suppose f is computable by a program called *halt* which has one input variable, say X, and one output variable, Y. If X reads the number of a program which halts on its own number, then *halt* writes 1 and then halts; otherwise, *halt* writes 0, and then halts.

By removing the read and write statements and by replacing them with suitable PASCAL syntax for passing parameters, we can regard *halt* as a function subprogram. We use it to construct a new program, *confuse*, which defines a partial function $\psi : \mathbf{N} \rightarrow \mathbf{N}$ as follows:

$$\psi(n) = \begin{cases} 1, & \text{if } halt(n) = 0; \\ \text{undefined}, & \text{otherwise.} \end{cases}$$

Here is *confuse*:

```
begin
    read(N);
    if halt(N) = 0
        then Y := 1
        else begin Y := 1;
            while Y ≠ 0 do
                Y := Y + 1
            end;
    write(Y)
end
```

If *halt* is a legal PASCAL program then so is *confuse*. Hence *confuse* has an index—obtainable directly from its ASCII representation. For definiteness, call that index e. How does ψ, the function computed by *confuse*, behave on input e?

$$\psi(e) = \begin{cases} 1, & \text{if } halt(e) = 0, \text{ i.e., if } confuse = P_e \text{ does not halt on } e; \\ \text{undefined}, & \text{if } halt(e) = 1, \text{ i.e., if } confuse = P_e \text{ halts on } e. \end{cases}$$

We have a contradiction! If *confuse* halts on argument e, then $halt(e)$ must have been 0, meaning that program e—*confuse*—on argument e never halts —a contradiction. On the other hand if *confuse* does not halt on argument e, then $halt(e)$ must have been 1. But then *confuse* must have halted on argument e—another contradiction. Hence program *confuse* cannot exist, and therefore program *halt* cannot exist either, proving our theorem: *No computer program can decide correctly if an arbitrary computer program halts on its own index*. Alternative statements for this result are: It is *undecidable* (or *unsolvable* or *recursively unsolvable*) whether an arbitrary computer program halts on its own index.

Strictly speaking, what we have shown is the nonexistence of *confuse* and *halt* as programs in an appropriately modified version of PASCAL. This is not, however, a pathology of the PASCAL fragment used here. The unsolvability of the Halting Problem in full generality will imply that *confuse* and *halt* cannot exist as programs in any programming language whatsoever. The proof of this theorem is often called a diagonal argument, because the counterexample program, *confuse*, is constructed from the behavior of each program on its own index. In the next section we consider this diagonalization process in more detail.

EXERCISES FOR SECTION 1.2

1. Let $\mathscr{Q}$ denote the set of capital Roman letters $\{A, B, \ldots, Z\}$ and x and y be arbitrary finite strings of letters from $\mathscr{Q}$—respectively called the *text string* and the *pattern*.

   (a) Describe a precise algorithm to determine whether a pattern y occurs in a text string x.
   (b) Describe a precise algorithm to determine the number of times a pattern y occurs in a text string x.

2. Consider the nth *power function* $\mathbf{N} \to \mathbf{N}$ which maps x to x^n, for a given positive integer n. We want to show that the nth power function is computable—or programmable in terms of the multiplication operation.

   (a) Describe an algorithm that computes the nth power function.
   (b) Describe a better algorithm that computes the nth power function. (For example, if $n = 31$ we can compute x^{31} according to the following sequence: $x, x^2, x^4, x^8, x^{16}, x^{24}, x^{28}, x^{30}, x^{31}$—which requires eight multiplications instead of 30.) Can you improve this algorithm if the division operation as well as multiplication are allowed as elementary operations?

3. A *Diophantine equation* is an equation of the form $P = 0$, where P is a polynomial with integer coefficients. For example,

   $$x^3 + 2xy + x^2 - y - 9 = 0$$

   is a Diophantine equation of degree 3 and in two variables, x and y.

   The general Diophantine Problem consists in finding *an algorithm to determine whether an arbitrary Diophantine equation has an integer solution.*[6] For the particular Diophantine equation mentioned above, the pair $x = 2$ and $y = -1$ is an integer solution.

   Give an algorithm—in flow-diagram form or as a sequence of precisely written instructions in English—for each of the following special cases of the

[6] The Diophantine Problem is also called Hilbert's Tenth Problem, after the German mathematician David Hilbert who first posed it in 1901 among a list of 20 outstanding unsolved mathematical problems. It took nearly 70 years to discover that there can be no algorithm for the general Diophantine Problem [Matijasevic (1970), Matijasevic and Robinson (1975)]; that is, the general Diophantine Problem is recursively unsolvable.

general Diophantine Problem:

(a) Given *positive* integers a, b, and c, are there *positive* integers x and y such that $ax^2 + by - c = 0$?
(b) Given integer coefficients a, b, c, and d, are there integers x, y, and z such that $ax + by + cz + d = 0$?
(c) Given integer coefficients $a_n, a_{n-1}, \ldots, a_0$, is there an integer x such that $a_n x^n + a_{n-1} x^{n-1} + \cdots + a_1 x + a_0 = 0$? (*Hint*: If x is a solution, then it must divide a_0.)

1.3 Diagonalization and the Halting Problem

The German mathematician, George Cantor, the inventor of modern set theory, was the first investigator to distinguish formally between the "sizes", or *cardinalities*, of two infinite sets. (For more background on the material in this section, see Section 5.3 of *A Basis for Theoretical Computer Science*.)

Recall that a total function $f: A \to B$ is *one-to-one*, or an *injection*, if for each $a_1 \neq a_2$ in A, we have $f(a_1) \neq f(a_2)$. A total function f is *onto*, or a *surjection*, if the range $f(A)$ equals all of B. A map is a *bijection* if it is both one-to-one and onto.

One set, A, is larger than or equal to another set, B, if there is an injection $f: B \to A$. If there is also an injection $g: A \to B$, then A is larger than or equal to B and B is larger than or equal to A. Cantor showed that under these circumstances there must be a bijection $h: A \to B$, and this is his definition of "same size as", or cardinality. Using this definition, are the set $\mathbf{N}$ of natural numbers and the set $\mathbf{R}$ of real numbers the same size? Cantor showed this wasn't so, using what is now called the *Cantor diagonalization argument*, which we present below.

1 Theorem (Cantor). *There is no bijection $h: \mathbf{N} \to \mathbf{R}$.*

PROOF. Suppose, by way of contradiction, that a bijection $h: \mathbf{N} \to \mathbf{R}$ does exist. That is, suppose the list

$$h(0), h(1), \ldots, h(n), \ldots$$

includes all of $\mathbf{R}$, and is without duplicates. We show how to construct a real number, r, that cannot appear in the list, thus contradicting the claim that h is a bijection. The real number r is constructed as follows: Imagine that each $h(k)$ is written in decimal notation—for example, $h(15)$ might be 126.34534967349 . . . . Our counterexample r will begin with a decimal point; the kth digit to the right of the decimal point will be a 5, unless the kth digit to the right of the decimal point of $h(k)$ is a 5, in which case the kth digit to the right of the decimal point of r is a 3. The number r so constructed differs from every real number in the list by at least one digit to

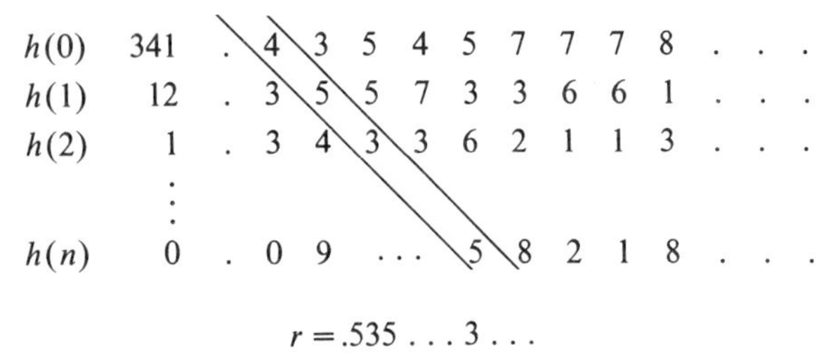

Figure 3 Construction by diagonalization of the missing real.

the right of the decimal point. The crucial digits come from the diagonal of the number grid representing our list of reals in Figure 3. □

Cantor's argument adapted to computability theory is an important technique for constructing noncomputable functions. To see this, we show how the argument carries over to the unsolvability of the Halting Problem. Figure 4 represents an infinite chart of all PASCAL programs (though this is not the chart for the listing in Section 1.2) on all inputs. Down arrows mean "halts," while up arrows mean "fails to halt."

The diagonal arrows are used in the definition of *confuse*. The unsolvability of the Halting Problem means that the chart in Figure 4 is not accessible to any algorithm: No computer program in PASCAL—or any other language—could calculate the direction of every arrow in the chart. For if a program could perform this calculation, even for the diagonal set of arrows, then a program could be constructed—*confuse*—which would differ in output behavior from every program in the list. But we know that this cannot happen.

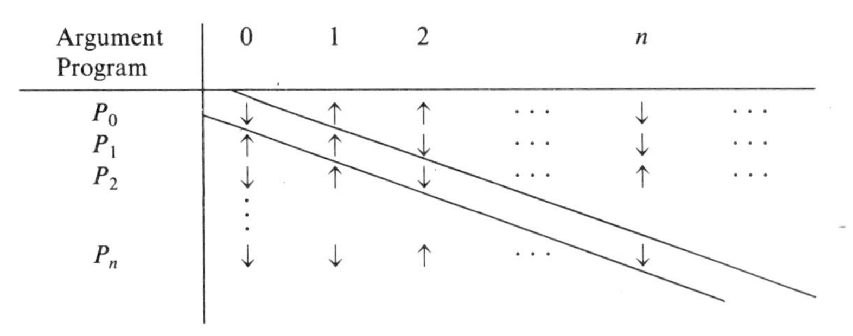

confuse has chart ↑↓↑ . . . ↑ . . .

Figure 4 The halting chart for PASCAL.

EXERCISES FOR SECTION 1.3

1. Show that each of the following sets is countably infinite, i.e., has the cardinality of **N**.

   (a) The Cartesian product $\mathbf{N} \times \mathbf{N}$.
   (b) The set of finite strings of letters from the alphabet $\{A, B, \ldots, Z\}$.

(c) The set of all (finite) PASCAL programs.
(d) The set of all rational numbers q such that $0 \leqslant q \leqslant 1$.
(e) The set of partial functions from $\mathbf{N}$ to $\mathbf{N}$ whose domains of definition are each finite.

2. Prove that the following sets are uncountable, i.e., have cardinality larger than that of $\mathbf{N}$.

(a) The set of all total functions from $\mathbf{N}$ to $\{0, 1\}$.
(b) The power set of $\mathbf{N}$, i.e., the set of all subsets of $\mathbf{N}$.
(c) The set of partial functions from $\mathbf{N}$ to $\mathbf{N}$ whose ranges are each finite.

3. We already know that there are number-theoretic functions which are not computable, the "halting" function f of Section 1.2 being one such function. Using a cardinality argument, establish this fact again. (*Hint*: There are more functions from $\mathbf{N}$ to $\mathbf{N}$ than there are programs for them.)

4. (Rogers) Consider the following function

$$f(x) = \begin{cases} 1, & \text{if there exists a run of at least } x \\ & \text{consecutive 5's in the decimal expansion of } \pi; \\ 0, & \text{otherwise.} \end{cases}$$

For example, $f(0) = f(1) = 1$, since $\pi = 3.1415926\ldots$. Is f computable? (*Hint*: What are the possible behaviors of f?)

5. Show that our proof of the undecidability of the Halting Problem does not hold if we assume that all our PASCAL programs are to be run on a specific general-purpose computer which has a core memory of size, say, 32 K.

6. Say that a real number is *computable* if there exists a PASCAL program which, on input $n \in \mathbf{N}$, outputs the first n places in the decimal expansion of the number.

(a) Show that every rational number is computable.
(b) Show that the square root of an integer is computable.
(c) Construct a "diagonal" argument to show that not every real number is computable.
(d) By considering the McLaurin expansion of the arctangent function and using a standard fact about the rate of convergence of an alternating series, it is possible to prove

$$\pi = 4 - 4/3 + 4/5 - 4/7 + 4/9 - \cdots$$

and that the sum of the first m terms is accurate to n decimal places if $m > 4 \times 10^{n+1}$. Assuming these results, prove that π is computable.
(e) Exactly one of the following three cases must be true:

Case 1. Four consecutive 5's do not occur in the decimal expansion of π.
Case 2. Case 1 is false and the smallest K with places $K, K+1, K+2, K+3$ each equal to 5 is an even number.
Case 3. As in case 2, except that K is an odd number.

Define two real numbers x, y as follows. If case 1 is true define $x = 0.33333\ldots$ and $y = 0.66666\ldots$. If case 2 is true define $x = 0.33333\ldots333444444\ldots$ with the change occurring in the Kth place and define $y = 0.66666\ldots$. If case 3 is true define $x = 0.33333\ldots$ and $y = 0.66666\ldots 66655555\ldots$ with the change occurring in the Kth place. Here is the punch line: Using the result of (d), prove that both x and y are computable. (*Hint*: Your algorithm need not know which of the three cases holds.) Show that the *first* decimal place of $x + y$ depends on which case holds. (It is a genuinely open question as to which case does hold. Hence, at the present time, it is possible to have a computer print out x and y to any number of decimal places, but it is not possible to print out even one decimal place of $x + y$!)

CHAPTER 2

The Syntax and Semantics of **while**-Programs

In Chapter 1 we presented a version of the unsolvability of the Halting Program for PASCAL. The generality of that result rests on the crucial assumption—embodied in our formulation of Church's Thesis—that the notion of "algorithmic" coincides with the notion of "programmable". In this chapter we examine the relationship between algorithms and programs more carefully. We define a particularly simple language, a PASCAL subset called **while**-programs, with the intent of isolating a set of primitive operations and combination rules which form a convenient and tractable basis for our formulation of computability theory.

2.1 The Language of **while**-Programs

All specifications of algorithms and corresponding characterizations of the algorithmic functions must begin with the assumption that certain elementary operations are inherently algorithmic. Indeed, without an initial set of operations that are given "for free", and whose algorithmic nature is never to be doubted, no characterization of the algorithmic functions can be undertaken. In our formulation of **while**-programs we take four basic operations "for free". The first allows us to write zero; the second, to add 1 to any natural number; the third, to subtract 1 from any natural number; and the fourth, to compare two arbitrary natural numbers and decide whether or not they are equal. It goes without saying that the less we take "for free", the less we have to argue in favor of our choice of basic operations, and the more "intuitively clear" our attempt will be to capture the notion of what is effective and computable.

Just as a specification of algorithms must start by assuming that some basic operations are inherently algorithmic, so too must it require that certain elementary rules for combining these basic operations (and other operations already built up from them) be allowed without further justification. There can be no realistic characterization of computability if no rules of combination are allowed, because under such an assumption the computable functions would then be limited to the few initial operations.

Thus our approach to the definition of a system of algorithmic specification is a special case of an *inductive definition*. Starting from a *basis*, we build up the set by repeated applications of the *inductive step* [Arbib, Kfoury, and Moll (1981), Chapter 2]. For example, we may regard the set **N** of non-negative integers as built from the basis whose only element is 0, by repeated application of the successor function. In our programming language we have two rules of combination. (Other specifications of algorithms include similar rules.) One rule specifies that the repeated execution of an action, already known to be algorithmic, an unbounded but finite number of times is itself algorithmic if the "stopping test" is algorithmic. The other rule says that the sequential execution of a specified finite sequence of algorithmic actions is itself an algorithmic action. Here, too, if we are parsimonious in choosing rules of combination, it is in order to increase the "credibility" of our characterization of computability.

Indeed, while our goal is a characterization of exactly the computable functions, if we fail in the attempt we would like this to be so because we have characterized *less* rather than *more* than all the computable functions.

The language of **while**-programs will therefore be a lean one compared to full high-level programming languages such as PASCAL or FORTRAN. But this is not really a shortcoming. For once we prove that the set of **while**-programs is fully as powerful as any computer language, we shall see that its simplicity is an important asset. This simplicity of description will help clarify the theoretical aspects of our work, and will make such goals as the construction of machinery for proving program correctness easier to achieve.

Let us add a few words about the execution of **while**-programs. Although they are not actually designed to run on a computer, they can in principle be executed by all instruction-obeying agents, such as digital computers. But because **while**-program execution is purely theoretical we can relax an important limitation of all real computers: We place no finite bound on computation time available during **while**-program processing. Thus if we prove that a function cannot be computed by any **while**-program, we are certainly guaranteeing that it cannot be computed by any program operating within time and memory limitations.

ASSIGNMENT STATEMENTS, **while**-STATEMENTS, COMPOUND STATEMENTS

The basic character set of our programming language contains:

1. *Variable names* which can each be an arbitrary finite string of upper case letters and decimal numerals provided it also starts with an upper case letter, e.g., $X1$, $X2$, NEW, TEMP, LST15, LST25, etc., but not 33 or 7AB13.
2. *Operator symbols* to denote the basic operators which change the values of the variables, namely *succ* for the successor function, *pred* for the predecessor function, and 0 for the constant zero function.
3. A *relation symbol* $\neq$ for the inequality which compares the values of pairs of variables.
4. *Program symbols* which are := (assignment symbol), ; (semi-colon), **begin**, **end**, **while**, and **do**.

We define a **while**-program to be a finite (possibly empty) sequence of *statements*, separated by semicolons, and preceded by **begin** and followed by **end**:

$$\textbf{begin } \mathbb{S}_1; \mathbb{S}_2; \ldots ; \mathbb{S}_n \textbf{ end}$$

where $\mathbb{S}_1, \mathbb{S}_2, \ldots, \mathbb{S}_n$ are arbitrary statements and $n \geqslant 0$.

Statements are defined inductively. The "assignment statements" form the basis step, while the induction step involves the formation of "**while** statements" and "compound statements".

In **while**-programs, the values assigned to variables will be non-negative integers. An *assignment statement* has one of the following forms:

1. $X := 0$
2. $X := succ(Y)$
3. $X := pred(Y)$

where *succ* is the "successor" function and *pred* is the "predecessor" function. [Recall that $pred(n) = n - 1$ if $n > 0$, and $pred(0) = 0$.]

X and Y in the above instructions stand for arbitrary variable names, possibly identical. In the diagrammatic representation of a **while**-program, an assignment statement will have one of the following forms:

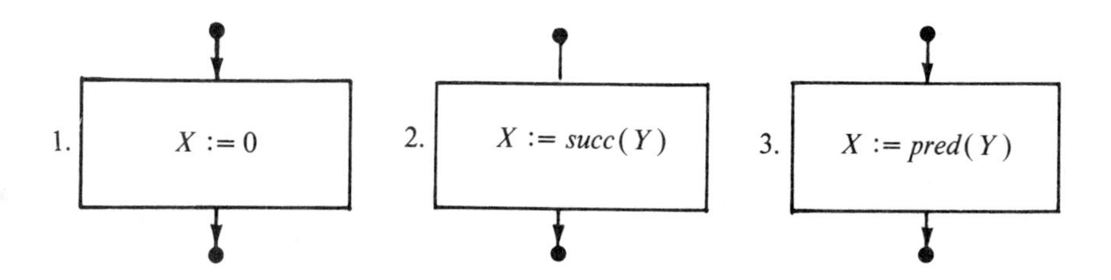

A **while** *statement* has the following form:

> **while** $X \neq Y$ **do** $\mathbb{S}$, which means "while the value of variable X is not equal to the value of variable Y, repeat the action specified by statement $\mathbb{S}$".

Diagrammatically, a **while** statement will have the following form:

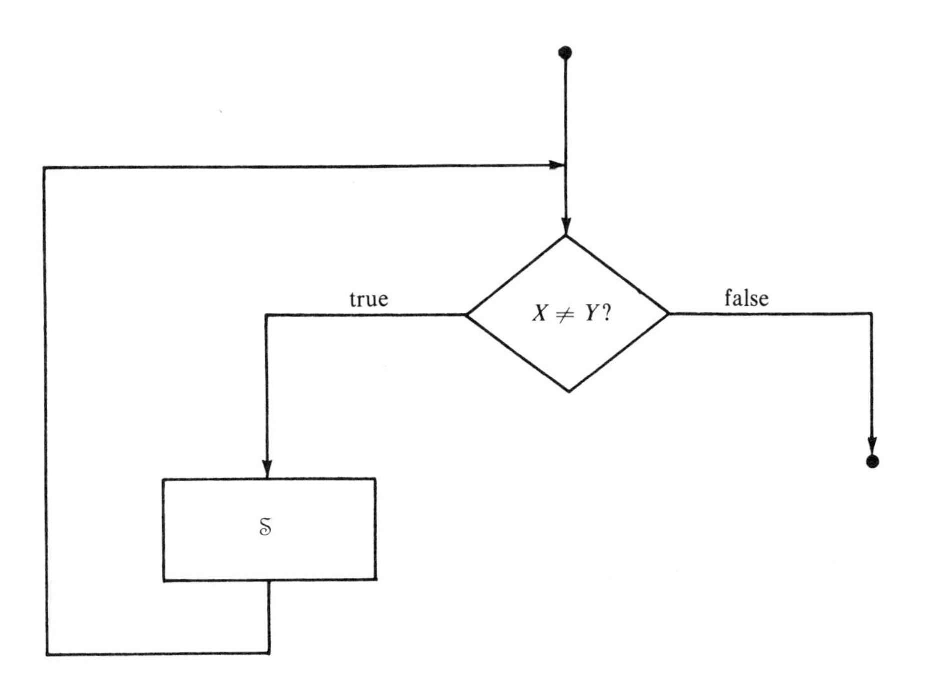

We use diamonds for tests, and rectangles for statements. $\mathbb{S}$ may be an arbitrary statement—assignment, **while**, or compound. We call $\mathbb{S}$ the *body* of the **while** statement and "$X \neq Y$" its *test*.

Finally, a *compound statement* has the form:

$$\textbf{begin } \mathbb{S}_1; \mathbb{S}_2; \ldots ; \mathbb{S}_m \textbf{ end}$$

where $\mathbb{S}_1, \mathbb{S}_2, \ldots, \mathbb{S}_m$ are arbitrary statements and $m \geqslant 0$. A compound statement such as the one given here simply means that we have to carry out the actions prescribed by $\mathbb{S}_1, \mathbb{S}_2, \ldots,$ and $\mathbb{S}_m$, in succession and in the given order. Diagrammatically, a compound statement will be represented by:

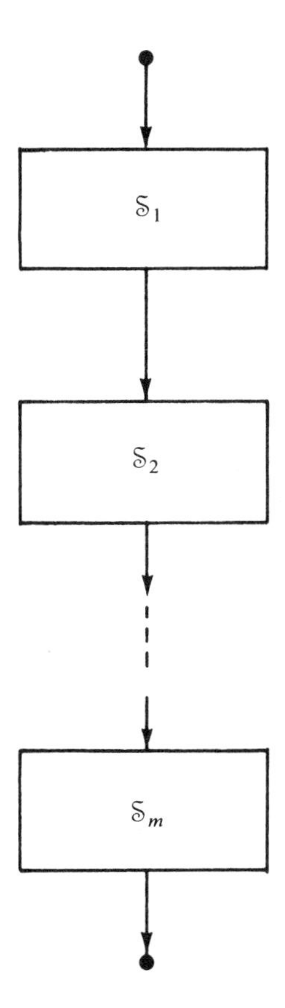

The case $m = 0$ is represented by the "straight-through" diagram, which corresponds to the computation which leaves the values of all variables unchanged.

Note that a **while**-program is none other than a compound statement which we choose to identify as a program, but which can also be used as a compound statement in some other larger program.

We see that each program eventually reduces to a "flow diagram" made up of assignment statements ($X := 0$, $X := succ(Y)$, or $X := pred(Y)$) and tests ($X \neq Y$?). We refer to each of these as an *instruction*. Thus, a **while** statement includes a test instruction in addition to the zero or more instructions in its body.

Notice that the flow diagram of a **while**-program has exactly one entry point and one exit point. This convention allows us to talk about the *first instruction* in a (nonempty) **while**-program and the *last instruction* before its exit point.

We now know how to write a "syntactically" correct **while**-program: We say that a **while**-program is *syntactically correct* if it satisfies the following inductive definition.

SUMMARY OF THE SYNTAX OF **while**-PROGRAMS

- Assignment statements:

$$X := 0, \qquad X := succ(Y), \qquad X := pred(Y).$$

where X and Y are arbitrary variable names.

- **while** statements:

$$\textbf{while } X \neq Y \textbf{ do } \mathbb{S}$$

where $\mathbb{S}$ is an arbitrary statement.

- Compound statements:

$$\textbf{begin } \mathbb{S}_1; \mathbb{S}_2; \ldots ; \mathbb{S}_m \textbf{ end}$$

where $\mathbb{S}_1, \mathbb{S}_2, \ldots, \mathbb{S}_m$ are arbitrary statements and $m \geqslant 0$.

- **while**-programs:

a **while**-program is a compound statement.

A GRAMMAR FOR **while**-PROGRAMS

We now give a more formal, grammatical account of the syntax of **while**-programs. This subsection may be omitted by readers unfamiliar with context-free grammars. (The elementary definitions for context-free grammars are given in Section 9.3.) We show that the set of **while**-programs is a context-free language by giving a set of productions which generate the set. We have a set ⟨digit⟩ of digits, and a set ⟨alpha⟩ of letters, so that variables are defined by the production:

$$\langle \text{variable} \rangle ::= \langle \text{alpha} \rangle \mid \langle \text{variable} \rangle \langle \text{alpha} \rangle \mid \langle \text{variable} \rangle \langle \text{digit} \rangle$$

The set of all assignment statements is given by the production:

$$\begin{aligned} \langle \text{asgt} \rangle ::= {} & \langle \text{variable} \rangle := 0 \mid \langle \text{variable} \rangle := succ(\langle \text{variable} \rangle) \mid \\ & \langle \text{variable} \rangle := pred(\langle \text{variable} \rangle) \end{aligned}$$

We define the set of tests by the production:

$$\langle \text{test} \rangle ::= \langle \text{variable} \rangle \neq \langle \text{variable} \rangle$$

Then statements may be defined by the production:

⟨statement⟩ : := ⟨asgt⟩ | **while** ⟨test⟩ **do** ⟨statement⟩ | ⟨program⟩

where a program is just a compound statement, defined by the productions:

⟨statement sequence⟩ : := ⟨statement⟩ | ⟨statement sequence⟩;
⟨statement⟩

⟨program⟩ : := **begin end** | **begin** ⟨statement sequence⟩ **end**

The reader should readily see that the set of terminal strings produced from the start-symbol ⟨program⟩ is indeed our set of **while**-programs.

Readers familiar with PASCAL and other languages of the ALGOL family will have recognized that the language of **while**-programs, as presented above, is a (small) fragment of such a language. In fact, many of the standard and useful features of ordinary computer languages are missing here—such as arithmetical operators (addition, multiplication, etc.), **goto** statements, **if-then-else** statements, and others. Our choice for a "lean" programming language was in part explained earlier; but, as we shall next see, this was not (and had better not be!) at the cost of being unable to program as many functions.

Exercises for Section 2.1

1. **while**-programs which do not use **while** statements are commonly called *straight-line programs*, since their flow diagrams contain no loops and appear as straight sequences of assignment statements.

   (a) Show that for every **while**-program with exactly one variable there is a straight-line program that computes the same number-theoretic function. As in Section 1.2, assume that programs receive input through a single read statement, and transmit output through a single write statement.
   (b) Exhibit a function computed by a **while**-program with exactly two variables, which is not computable by any straight-line program. (*Hint*: First prove that functions computed by straight-line programs are total.)

2. We define the class of *flowchart programs* as follows. Flowchart programs are constructed from "assignment statements" and "conditional statements". As in the case of **while**-programs, an *assignment statement* has one of the following forms:

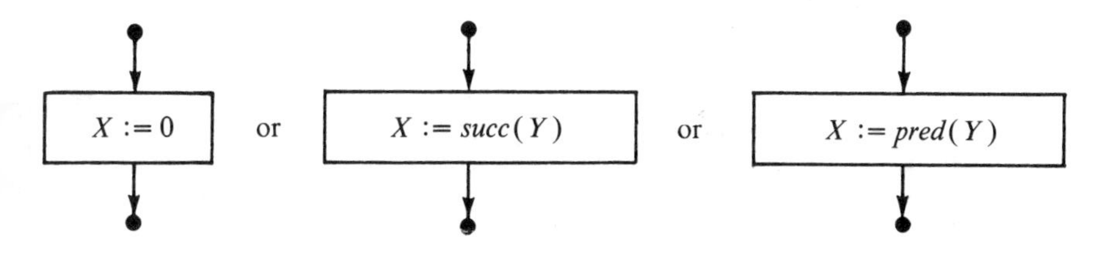

where X and Y are arbitrary variable names. A *conditional statement* is of the form:

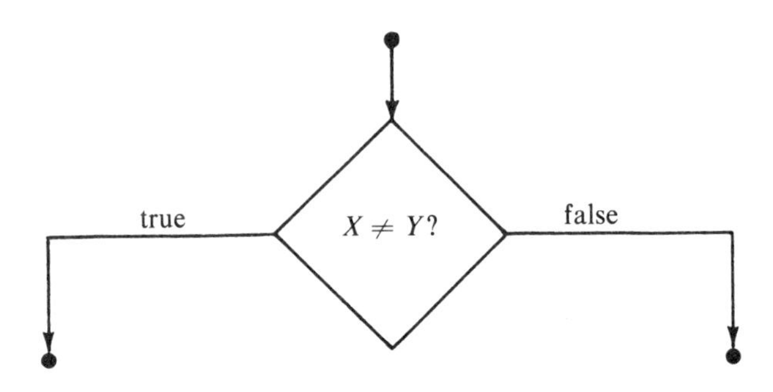

A flowchart program is simply *any* finite flow diagram built up from assignment and conditional statements, with exactly one entry point and finitely many exit points (possibly no exit points, since every outgoing edge may be made to re-enter a statement in the flowchart).

(a) Show that (the flow diagram of) every **while**-program is a flowchart program.
(b) Show that a flowchart program of the form shown below cannot be the flow diagram of a **while**-program.

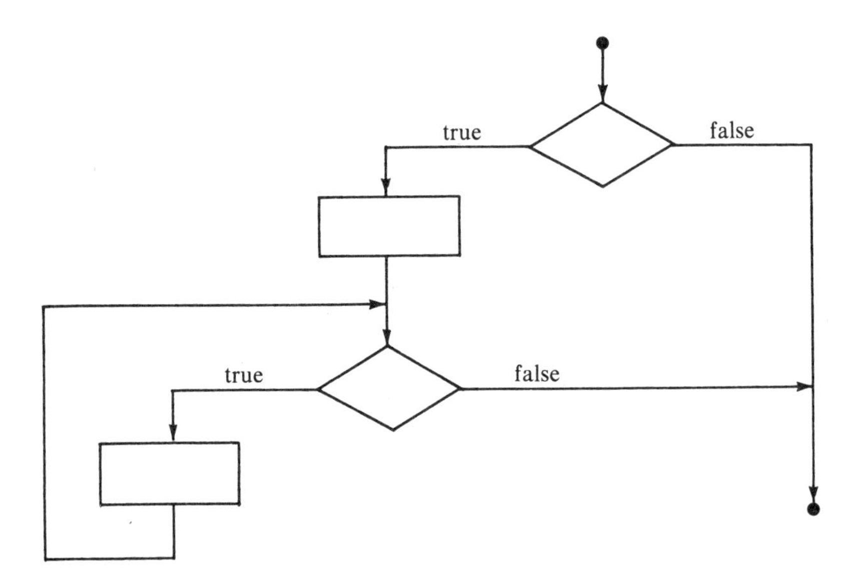

(c) Fill in the "square" and "diamond" boxes in the flowchart of part (b) in some arbitrary way, and then find a **while**-program which is equivalent to the resulting flowchart program. As in Section 1.2, assume that all programs receive input through a single read statement inserted just after the entry point, and transmit output through a single write statement inserted before every exit point.

3. A **goto**-program is a **while**-program where "**goto** statements" and "labels" are allowed. A **goto** *statement* is of the form

goto n

where n is a natural number. A *label* is a natural number followed by a colon and placed before a statement in a program. Executing a **goto** statement, say "**goto** n", causes statement with label n to be executed next. For example, if we have the following context in a program:

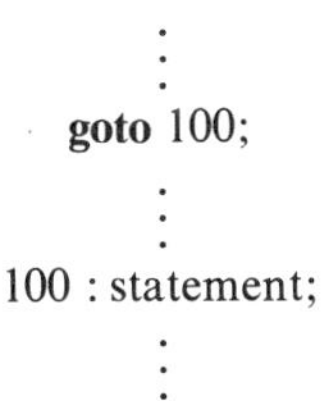

⋮
goto 100;
⋮
100 : statement;
⋮

execution of **goto** 100 will result in the statement with label 100 being executed next.

In the case of **goto**-programs two additional syntactic rules are introduced. A **goto**-program is syntactically correct only if (1) every label mentioned in a **goto** statement is the label of some statement in the program, and (2) every label is the label of at most one statement in the program.

Prove that the class of flowchart programs, as defined in the preceding exercise, and the class of **goto**-programs are intertranslatable. That is, show that the flow diagram of every **goto**-program is a flowchart program, and every flow chart program is the flow diagram of some **goto**-program.

2.2 Macro Statements

We now show how to write a **while**-program P that adds X to Y and leave the resulting value in Z. (By a slight abuse of language, we shall not always distinguish, here and in the sequel, between a variable and the value assigned to it.)

The operation to be performed by P may be decomposed as follows:

1. Assign the value of X to Z.
2. Add the value of Y to Z.

Concerning step 1 there is no single statement in our programming language that will carry it out. It is easily seen, however, that the following compound statement will implement step 1.

begin $Z := succ(X)$; $Z := pred(Z)$ **end**

We may summarize this compound statement by the expression $Z := X$, which we shall call a *macro statement* to indicate that it is an abbreviation for a multi-instruction statement in our basic programming language. And whenever we use a macro statement in a **while**-program, we shall understand that it is a short-hand notation for its *macro definition*, the corresponding compound statement.

Likewise, step 2 in the operation of P cannot be implemented directly in the language of **while**-programs. It too will correspond to a macro statement. Let us describe this macro by first rewriting step 2 as

2. Add 1 to Z, Y times.

Step 2 may be decomposed further using an auxiliary variable U, as

2. (i) Set U to 0.
2. (ii) Add 1 to Z and add 1 to U, as many times as required to make $U = Y$.

Now the implementation of step 2 is clear:

```
begin
    U := 0;
    while U ≠ Y do
    begin Z := succ(Z); U := succ(U) end
end
```

This implementation may be summarized by the following macro statement,

$$Z := Z + Y.$$

Finally, writing program P in terms of the two previously defined macro statements, we have

```
begin
    Z := X;
    Z := Z + Y
end
```

which itself may be summarized by a new macro, namely $Z := X + Y$.

Whenever we use the macro statement $Z := X + Y$ in a **while**-program, we shall understand that it abbreviates the following compound statement (the program P)

```
begin
    begin Z := succ(X); Z := pred(Z) end;
        begin U := 0;
            while U ≠ Y do
            begin Z := succ(Z); U := succ(U) end
        end
end
```

which we shall call the *macro definition* of $Z := X + Y$. Note that arguments X and Y in the macro statement $Z := X + Y$ do not appear on the left-hand side of := in the corresponding macro definition (the program P). This guarantees that X and Y still hold their initial values after execution of the macro statement. Furthermore, to avoid problems resulting from giving the same name to different variables, we assume that auxiliary variable U appears in P only, and appears nowhere in a program using macro statement $Z := X + Y$. This is always possible since we have an infinite supply of variables.

Macro statements for useful language features

The statement $Z := X + Y$ in the example above is called "macro" because it is not part of the repertoire of basic statements in our programming language, but is really a short-hand notation for some more elaborate compound statement (its so-called macro definition).

Our motivation for further examining the topic of macro statements is two-fold—first theoretical, and second practical. First, we want to show that any statement encountered in practice in our programming languages is equivalent to a macro statement in the language of **while**-programs. This will allow us to conclude that the expressive power of any known programming language is no more than that of **while**-programs. It will be the first piece of evidence we present in the book to support Church's Thesis. (We shall limit our aim to showing that the language of **while**-programs lets us handle the essential features of one language that is widely used, namely, PASCAL, and then let the reader discover through practice or other means that any numerical computation programmable in any other language is also programmable in PASCAL.)

Second—this is the practical side—it becomes a tedious and onerous task to write all **while**-programs in terms of only the basic statements in our programming language. It is therefore a considerable saving of effort if we allow the use of macro statements to substitute for relatively long and commonly used compound statements.

For the next proposition, let us agree on the following notation: $*$ stand for multiplication, $**$ for exponentiation (actually not present in all versions of PASCAL, although programmable in it), **div** for integer division, **mod** for remainder of integer division, and $\dot{-}$ for monus (also not present in PASCAL). For our purposes binary operators are defined only over the natural numbers. The integer division $x\,\mathbf{div}\,y$ has for value the integer quotient obtained on dividing x by y, and $x\,\mathbf{mod}\,y$ the remainder obtained on dividing x by y so that

$$x = y * (x\,\mathbf{div}\,y) + x\,\mathbf{mod}\,y.$$

We agree to set $x\,\mathbf{div}\,0 = 0$ and $x\,\mathbf{mod}\,0 = x$, for all $x \in \mathbf{N}$. As for $x \dot{-} y$, it is the following operation

$$x \dot{-} y = \begin{cases} x - y, \text{ if } x \geqslant y; \\ 0, \text{ otherwise.} \end{cases}$$

1 Proposition. *The following are macro statements in the language of* **while**-*programs*:

(a) $Y := X$,
(b) $X := n$, *where* n *is a given natural number*,
(c) $Z := X + Y$,
(d) $Z := X \dot{-} Y$,
(e) $Z := X * Y$,
(f) $Z := X \,\mathbf{div}\, Y$,
(g) $Z := X \,\mathbf{mod}\, Y$,
(h) $Z := X ** Y$,
(i) $Z := 2^X$,
(j) $Z := log_2(X)$,

i.e., each can be computed by a suitable **while**-*program.* (*Note that we here define* $log_2(X)$ *to be integer valued as the largest* y *such that* $2^y \leqslant X$: $log_2(15) = 3$, $log_2(16) = 4$.)

PROOF. It suffices to write the appropriate macro definitions. (a) and (c) were already shown, in the example above, to be equivalent to compound statements in the language of **while**-programs. For (b), we have the following macro definition

$$\mathbf{begin}\ X := 0; \underbrace{X := succ(X); \ldots; X := succ(X)}_{n \text{ times}}\ \mathbf{end}$$

For (d) we have (recalling that *pred* "locks" at 0)

```
begin
  Z := X; U := Y; V := 0;
  while U ≠ V do begin Z := pred(Z);
                       U := pred(U) end
end
```

Note that the first and second statements in this macro definition are themselves macro statements, which, by (a), have already been shown to be acceptable macro statements in the language of **while**-programs.

For (e), (f), (g), (h), (i), and (j), see Exercise 1. □

For further convenience, we shall allow ourselves to use composite macro statements of the form: $X := \mathcal{E}$, where $\mathcal{E}$ is an arithmetical expression written in terms of the operators introduced above. For example, we may legally write

$$X := succ(Y \dot{-} Z) + pred(Z)$$

which can be decomposed into the following sequence of statements:

begin
  TEMP1 := $pred(Z)$;
  TEMP2 := $Y \dot{-} Z$;
  TEMP2 := $succ$(TEMP2);
  X := TEMP2 + TEMP1
end

Test macros

The only tests we can use in **while** statements so far are of the form $X \neq Y$.

Other possible tests are of the form $X = Y$, $X < Y$, $3 < X$, $4 < 5$, etc., with their usual meanings.

From the preceding tests, we can construct compound tests that relate more than just two items. For this we use the following logical operators:

conjunction, written $\wedge$ ("and"),
disjunction, written $\vee$ ("or"),
negation, written $\neg$ ("not").

We define the set of *tests* inductively as follows:

Basis: If α and β are either variables or natural numbers, then each of $\alpha = \beta$, $\alpha \neq \beta$ and $\alpha < \beta$ is a test.
Induction Step: If $\mathcal{C}_1$ and $\mathcal{C}_2$ are tests, then so too are $(\mathcal{C}_1 \wedge \mathcal{C}_2)$, $(\mathcal{C}_1 \vee \mathcal{C}_2)$, and $\neg\mathcal{C}_1$.

There are useful laws to manipulate and simplify tests—the commutative, associative, and distributive laws, as well as De Morgan's laws. We shall not discuss these laws here but refer the reader to Section 4.1 of Arbib, Kfoury, and Moll (1981).

2 Proposition. *A statement of the form*

while $\mathcal{C}$ **do** $\mathcal{S}$,

where $\mathcal{C}$ is a test and $\mathcal{S}$ is an arbitrary statement, is a macro statement in the language of **while**-*programs.*

Proof. Without loss of generality we may assume that the test $\mathcal{C}$ does not involve natural numbers, and only relates variables. For if $\mathcal{C}$ were written in terms of natural numbers $n_1, \ldots, n_k$, which we may express by writing $\mathcal{C}(n_1, \ldots, n_k)$, we could replace the macro statement under consideration by the following equivalent sequence:

$N1 := n_1$;
⋮
$Nk := n_k$;
while $\mathcal{C}(N1, \ldots, Nk)$ **do** $\mathcal{S}$

where $\mathcal{C}(N1, \ldots, Nk)$ is now written only in terms of variables.

We now show that, given an arbitrary test $\mathcal{C}$, we can write an arithmetical expression $E_{\mathcal{C}}$ in terms of the variables used in $\mathcal{C}$ and the arithmetical operators *succ*, *pred*, $+$, and $\dot{-}$, such that $E_{\mathcal{C}} = 1$ just in case $\mathcal{C}$ is *true* and $E_{\mathcal{C}} = 0$ just in case $\mathcal{C}$ is *false*. Given that $+$ and $\dot{-}$ are programmable in the language of **while**-programs, as shown in the preceding proposition, this implies that the following sequence

```
begin
    U := E_C;
    V := 0;
    while U ≠ V do
        begin S; U := E_C end
end
```

where variables U and V are assumed not to appear elsewhere, is programmable in the language of **while**-programs and is equivalent to the statement: **while** $\mathcal{C}$ **do** $\mathcal{S}$.

It remains to show how to find $E_{\mathcal{C}}$ for an arbitrary test $\mathcal{C}$. We do this using the inductive definition of tests. Let us take tests of the form $X < Y$ to be the simplest kind or "atomic", and classify all other tests to be "composite". [We can consider $X \neq Y$ and $X = Y$ to be composite, because the first is equivalent to $(X < Y) \vee (Y < X)$ and the second is equivalent to $\neg(X \neq Y)$.]

Now, if $\mathcal{C}$ is atomic and of the form $X < Y$, then we let $E_{\mathcal{C}}$ be

$$(Y \dot{-} X) \dot{-} pred(Y \dot{-} X)$$

and it is readily seen that if $\mathcal{C}$ is *true*, $E_{\mathcal{C}} = 1$; and if $\mathcal{C}$ is *false*, $E_{\mathcal{C}} = 0$.

Assume we have found expressions $E_{\mathcal{C}_1}$ and $E_{\mathcal{C}_2}$ for tests $\mathcal{C}_1$ and $\mathcal{C}_2$. If $\mathcal{C}$ is $\mathcal{C}_1 \wedge \mathcal{C}_2$, we let $E_{\mathcal{C}}$ be

$$pred(E_{\mathcal{C}_1} + E_{\mathcal{C}_2}).$$

If $\mathcal{C}$ is $\mathcal{C}_1 \vee \mathcal{C}_2$, we let $E_{\mathcal{C}}$ be

$$(E_{\mathcal{C}_1} + E_{\mathcal{C}_2}) \dot{-} pred(E_{\mathcal{C}_1} + E_{\mathcal{C}_2}).$$

And if $\mathcal{C}$ is $\neg\mathcal{C}_1$, we let $E_{\mathcal{C}}$ be

$$1 \dot{-} E_{\mathcal{C}_1}.$$

In each of the three composite cases, it is easily verified that $\mathcal{C}$ is *true* if $E_{\mathcal{C}} = 1$, and *false* if $E_{\mathcal{C}} = 0$. □

There are "structured statements", not directly available in the language of **while**-programs, which are common features of other programming languages and quite useful in shortening the description of algorithms. These structured statements are of the following form, where $\mathcal{C}$ is an arbitrary test, and $\mathcal{S}_1$ and $\mathcal{S}_2$ arbitrary statements:

(1) **if** $\mathcal{C}$ **then** $\mathbb{S}_1$ **else** $\mathbb{S}_2$, which prescribes the following action: "evaluate test $\mathcal{C}$, and if it is *true* carry out $\mathbb{S}_1$, otherwise carry out $\mathbb{S}_2$". It is diagrammatically represented by

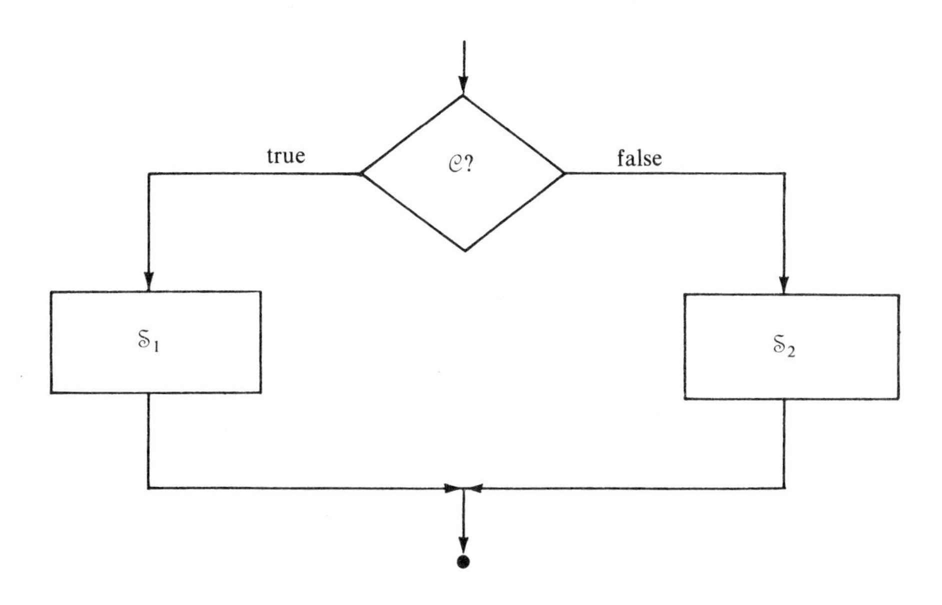

(2) **if** $\mathcal{C}$ **then** $\mathbb{S}_1$, which prescribes the following action: "evaluate $\mathcal{C}$, if it is *true* carry out $\mathbb{S}_1$, otherwise go directly to the next statement"—diagrammatically,

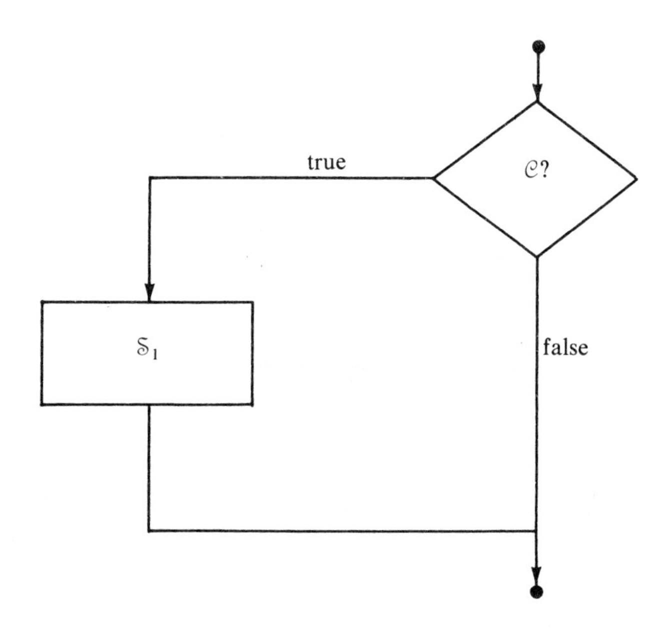

(3) **repeat** $\mathbb{S}_1$ **until** $\mathcal{C}$, which means "execute the action prescribed by $\mathbb{S}_1$ and then evaluate $\mathcal{C}$, if $\mathcal{C}$ is *false* then execute the action of $\mathbb{S}_1$ repeatedly until $\mathcal{C}$ becomes *true*"—diagrammatically,

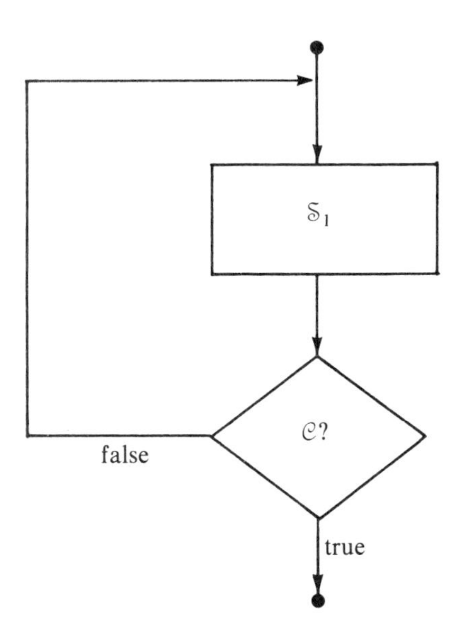

3 Proposition. *The* **if-then-else**, **if-then**, *and* **repeat-until** *statements, as defined above, are all macro statements in the language of* **while**-*programs.*

PROOF. We write an appropriate macro definition for **if** $\mathcal{C}$ **then** $\mathbb{S}$, leaving the other constructs for Exercise 2. Clearly, the first approximation to **if** $\mathcal{C}$ **then** $\mathbb{S}$ is: **while** $\mathcal{C}$ **do** $\mathbb{S}$—but if $\mathcal{C}$ is true, we wish to execute $\mathbb{S}$ only once, rather than repeatedly. We thus introduce an auxiliary variable V, and write the program

begin $V := 0$;
  while $\mathcal{C} \wedge (V = 0)$ **do**
  begin $\mathbb{S}$; $V := succ(V)$ **end**
end

We start by setting V to 0. If $\mathcal{C}$ is false we exit the **while-do** immediately anyway; but if $\mathcal{C}$ is true we can only do $\mathbb{S}$ once, for after the first execution of $\mathbb{S}$, V is increased to 1, and so the $V = 0$? test fails irrespective of the truth value of $\mathcal{C}$. □

With the macro statements introduced in the three preceding propositions, the extended language of **while**-programs includes most of the programming features used in ordinary computer languages in numerical computations. In fact, our extended language is essentially PASCAL without declarations (we just help ourselves to all the variables we need) and without its special features that are suited for handling data types other than non-negative integers.

Henceforth, whenever convenient, we shall use the extended language of **while**-programs which includes all the macro statements defined above. As we shall see, this will help us considerably in writing concise programs for relatively complex computations. At the same time we are entitled to revert to the basic language of **while**-programs whenever we deem it useful by replacing each macro statement by its macro definition. Indeed, defining concepts and proving properties about **while**-programs is a much easier task when the only statements we have to consider are the basic ones. An illustration of this fact will be provided in Section 2.3 when we discuss the operational semantics of **while**-programs.

Another important aspect of ordinary PASCAL (and programming languages in general) is the ability to write subprograms, sometimes defined recursively. We would also like to include this important programming feature in our PASCAL fragment. The details of this extension are briefly discussed in Section 5.2. However, we reserve the term **while**-programs for the nonrecursive programs defined in this and the previous section. Thus, in writing macros, **while**-programs *may call each other as subprograms so long as no self, i.e., recursive, call is introduced, even via a chain of intervening calls*. This ensures that we shall always be able to remove subprogram calls from our **while**-programs in finitely many steps, by replacing each call by the corresponding subprogram. In this way we may conclude that if a function is computed by a **while**-program with subprogram calls then it is also computed by one without subprogram calls. This finite elimination of subprogram calls is not possible with recursive programs.

Fortunately, this stipulation does not weaken the computational power of **while**-programs. We shall see in Section 5.2 that, given any program with recursive calls used for *numerical computations*, there is an equivalent program without recursive calls.[1]

EXERCISES FOR SECTION 2.2

1. Write **while**-programs for the following macro statements.

   (a) $Z := X * Y$,
   (b) $Z := X$ **div** Y,
   (c) $Z := X$ **mod** Y,
   (d) $Z := X ** Y$,
   (e) $Z := 2^X$,
   (f) $Z := log_2(X)$.

[1] This fact can also be deduced from results of Arbib (1963) and Shepherdson and Sturgis (1963). Parts of these two papers concerning us here are nicely summarized by Manna (1974) in the section on "Post Machines". For a theoretical discussion of what it is about numerical computations that makes recursive programs no more powerful than iterative programs, we refer the reader to Kfoury (1974). All these results confirm the experienced programmer's intuition that a language such as FORTRAN, which does not allow recursively defined programs, has as much "computational power" as programming languages with recursions.

For part (d) above, we agree to define $0 ** 0 = 0$; and for (f), we set $log_2(0) = 0$ (contrary to usual convention).

2. Write macro definitions for **if-then-else** and **repeat-until** constructions.

3. Write a **while**-program which computes $pred(X)$, using only assignment statements of the form $X := 0$ and $X := succ(X)$.

4. Show that replacing statements of the form **while** $X \neq Y$ **do** with **while** $X \neq 0$ **do** gives an equivalent formulation of **while**-programs.

5. We use the definition of flowchart programs and **goto**-programs given in Exercises 1.2 and 1.3.

   (a) Let P be an arbitrary **goto**-program with n basic instructions, i.e., the flow diagram of P has n "square" and "diamond" boxes. Transform P into an equivalent flowchart program P' where every basic instruction is given a unique label from the set $\{1, 2, \ldots, n\}$, and 1 is the label of the first instruction in P'.

   (b) Show that flowchart P' can be transformed into an equivalent **while**-program Q of the form:

```
begin
    COUNTER := 1;
    while COUNTER ⩽ n do
    begin if COUNTER = 1 then S_1;
          if COUNTER = 2 then S_2;
          .
          .
          .
          if COUNTER = n then S_n
    end
end
```

   where COUNTER is a new variable used as an "instruction counter", and $\mathbb{S}_i$ is a compound statement appropriately defined from the instruction with label i in P', $i \in \{1, 2, \ldots, n\}$. Conclude that **goto**-programs are no more powerful than **while**-programs.

   (c) A **while** statement W_1 is *nested* in a **while** statement W_2 if W_1 is in the body of W_2. A sequence of **while** statements $W_1, W_2, \ldots, W_n$ is said to be a *nested sequence of depth n* if W_i is nested in W_{i+1} for all $1 \leqslant i < n$. Use the construction of part (b) to show that any **goto**-program P (and therefore any **while**-program) can be transformed into an equivalent **while**-program Q where nesting of while statements is of depth $\leqslant 2$. (Note: Do not use macro statements in Q. If macros of the form **if-then** and **if-then-else** are allowed, Q does not need to use more than one **while** statement.)

2.3 The Computable Functions

While we have discussed the formal description and behavior of **while**-programs at length, we have not yet spelled out in explicit detail how **while**-programs may be thought of as objects which compute number-theoretic functions. And so in this section we describe the *semantics* of **while**-programs—a set of conventions which will permit us to associate a number-theoretic partial function with each program in our programming language.

Our approach in this section is called *operational*, in that it is based on the concept of a computation, the sequence of elementary instructions or *operations* executed when a program runs on a particular set of inputs. (We shall discuss a different approach, denotational semantics, in Section 5.1.) If the sequence never terminates, the program is undefined on those inputs; otherwise the program is defined, and the length of the sequence is a reasonable measure of the *complexity* of the computation.

COMPUTATIONS BY **while**-PROGRAMS

Although it may be intuitively clear how a **while**-program behaves on the basis of our informal explanation of the action of each instruction from the basic instruction set, we still need to give a precise definition for the notion of a **while**-program computation.

First of all, we recall that all permissible values of the variables are non-negative integers; no negative, real, complex, Boolean, or other values are permitted. Second, let us agree that whenever we specify a statement $\mathbb{S}$ (and thus in particular, a **while**-program P), we also specify a positive integer k, and call $\mathbb{S}$ a k-variable statement. This will mean that $\mathbb{S}$ uses *at most* k variables, but does not imply that it uses exactly k. We also adopt the convention that if $\mathbb{S}$ is a k-variable statement, then the variables it uses are (a subset of) $X1, X2, \ldots, Xk$. This convention assures us that if, for example, $\mathbb{S}_1$ and $\mathbb{S}_2$ are k-variable statements, then so too is their composite **begin** $\mathbb{S}_1$; $\mathbb{S}_2$ **end**, even though $\mathbb{S}_1$ may only involve, say, $X1$ and $X2$ while $\mathbb{S}_2$ involves $X2$ and $X3$.

Let us also agree to order $X1, \ldots, Xk$ according to their indices, from 1 to k, so that their values at any point in time may be specified by a vector of dimension k,

$$\mathbf{a} = (a_1, a_2, \ldots, a_k) \text{ in } \mathbf{N}^k$$

where a_i is the natural number assigned to variable Xi, for $1 \leqslant i \leqslant k$.

1 Definition. A *state of computation* for a k-variable **while**-program is a k-dimensional vector over the natural numbers. We shall also call it a *state vector*.

Consider the following two-variable program (which is equivalent to $X1 := X2$).

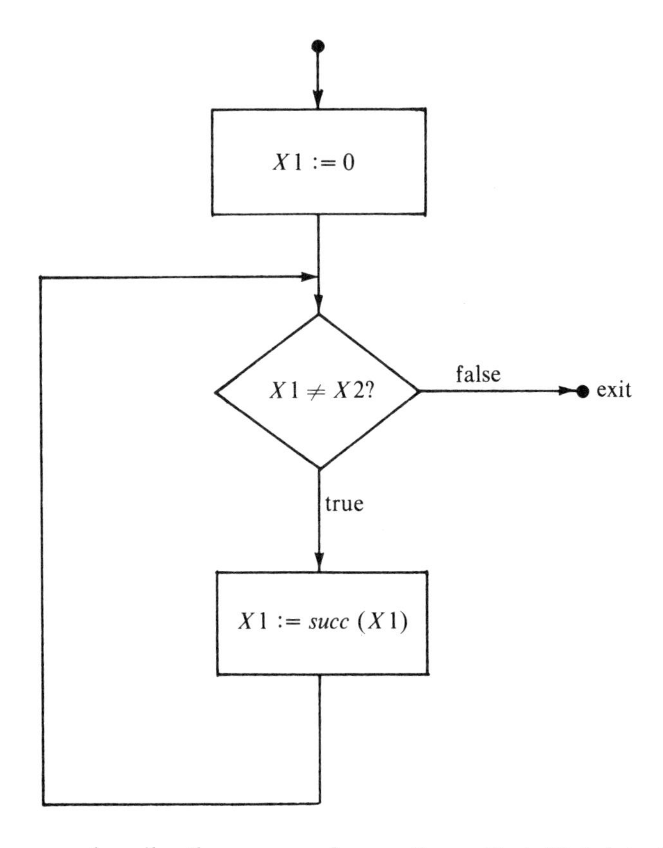

Then we may describe the course of execution with initial data $(X1, X2) = (4, 2)$ by the sequence

$$\begin{gathered}(4,2)\\ X1 := 0\\ (0,2)\\ X1 \neq X2?\\ (0,2)\\ X1 := succ(X1)\\ (1,2)\\ X1 \neq X2?\\ (1,2)\\ X1 := succ(X1)\\ (2,2)\\ X1 \neq X2?\\ (2,2)\end{gathered}$$

where each state vector is thought of as occurring at a particular location in the program, and is followed in turn by the next instruction and then by the state vector obtained by applying that instruction to the prior vector.

We thus consider the following general setting.

2 Definition. Let P be a k-variable **while**-program. A *computation* by P is a sequence, possibly infinite, of the form

$$\mathbf{a}_0 A_1 \mathbf{a}_1 A_2 \mathbf{a}_2 \ldots \mathbf{a}_i A_{i+1} \mathbf{a}_{i+1} \ldots,$$

where the $\mathbf{a}_i$'s are k-dimensional state vectors and the A_i's instructions appearing in P, which satisfies the consistency conditions below.

If the computation sequence is finite, then it has the form

$$\mathbf{a}_0 A_1 \mathbf{a}_1 A_2 \mathbf{a}_2 \ldots \mathbf{a}_{n-1} A_n \mathbf{a}_n, \qquad n \geqslant 0$$

i.e., its last term $\mathbf{a}_n$ is a state vector, and wc call n the *length* of the computation.

The consistency conditions for a computation by P are:

(1) There is a path through the flow diagram of P, starting with its entry point, such that the sequence of instructions appearing along this path is $A_1 A_2 \ldots A_i \ldots$. Here A_1 is necessarily the first instruction in P. If, in addition, the path is finite, we require that it terminates at the exit point of P, and the corresponding sequence $A_1, A_2, \ldots, A_n$ is such that A_n is the last instruction encountered before exit.

(2) The first term $\mathbf{a}_0$ in the sequence $\mathbf{a}_0 \mathbf{a}_1 \ldots \mathbf{a}_i \ldots$ is an arbitrary k-tuple of natural numbers. For all $i \geqslant 1$, if $\mathbf{a}_{i-1} = (a_1, a_2, \ldots, a_k) \in \mathbf{N}^k$ and A_i is a test instruction, then $\mathbf{a}_i$ is identical to $\mathbf{a}_{i-1}$; and if A_i is an assignment instruction of the form

$$Xu := g(Xv),$$

where $g(Xv)$ is either $succ(Xv)$ or $pred(Xv)$ or 0, and $1 \leqslant u, v \leqslant k$, then

$$\mathbf{a}_i = (a_1, \ldots, a_{u-1}, g(a_v), a_{u+1}, \ldots, a_k).$$

(3) In the sequence $A_1 A_2 \ldots A_i \ldots$, if A_i is an assignment instruction, then A_{i+1} is the next instruction after A_i in P if such an instruction exists; otherwise the sequence is finite and A_i is its last term. If A_i is a test instruction of the form

$$Xu \neq Xv?$$

where $1 \leqslant u, v \leqslant k$ and $\mathbf{a}_{i-1} = (a_1, \ldots, a_k) \in \mathbf{N}^k$ is such that $a_u \neq a_v$, then A_{i+1} is the first instruction within the body of the corresponding **while** statement; otherwise, if $a_u = a_v$, A_{i+1} is the first instruction after the **while** statement if such an instruction exists; otherwise the sequence is finite and A_i is its last term.

We clearly have the following:

3 Proposition. *Let P be a k-variable* **while**-*program and $\mathbf{a}_0$ an arbitrary k-tuple of natural numbers. Then there is exactly one computation by P that has $\mathbf{a}_0$ as its initial state vector.*

We are now entitled to talk about *the* computation by P that starts with initial state vector $\mathbf{a}_0$, which we shall also call an *input vector*.

4 Definition. If the computation by P which starts with $\mathbf{a}_0$ is infinite, we shall say that P *diverges* on input $\mathbf{a}_0$ and its output is undefined. If, on the other hand, the computation sequence is finite:

$$\mathbf{a}_0 A_1 \mathbf{a}_1 A_2 \mathbf{a}_2 \ldots \mathbf{a}_{n-1} A_n \mathbf{a}_n ,$$

we shall say that P *converges* or *terminates* on input $\mathbf{a}_0$ (in n steps) and its output is $\mathbf{a}_n$.

We stress that the concept of a computation by a **while**-program is defined here without mention of a computing machine, real or abstract.

5 Example. Consider the following **while**-program Q.

begin while $X1 \neq X2$ **do**
  begin $X1 := succ(X1)$; $X3 := pred(X3)$ **end**;
 while $X2 \neq X3$ **do** $X1 := 0$
end

The flow diagram of Q appears on page 39. The computation by Q with initial state vector

$$(X1, X2, X3) = (4, 6, 8)$$

is the following sequence:

$$(4,6,8)$$
$$X1 \neq X2?$$
$$(4,6,8)$$
$$X1 := succ(X1)$$
$$(5,6,8)$$
$$X3 := pred(X3)$$
$$(5,6,7)$$
$$X1 \neq X2?$$
$$(5,6,7)$$
$$X1 := succ(X1)$$
$$(6,6,7)$$
$$X3 := pred(X3)$$
$$(6,6,6)$$
$$X1 \neq X2?$$
$$(6,6,6)$$
$$X2 \neq X3?$$
$$(6,6,6)$$

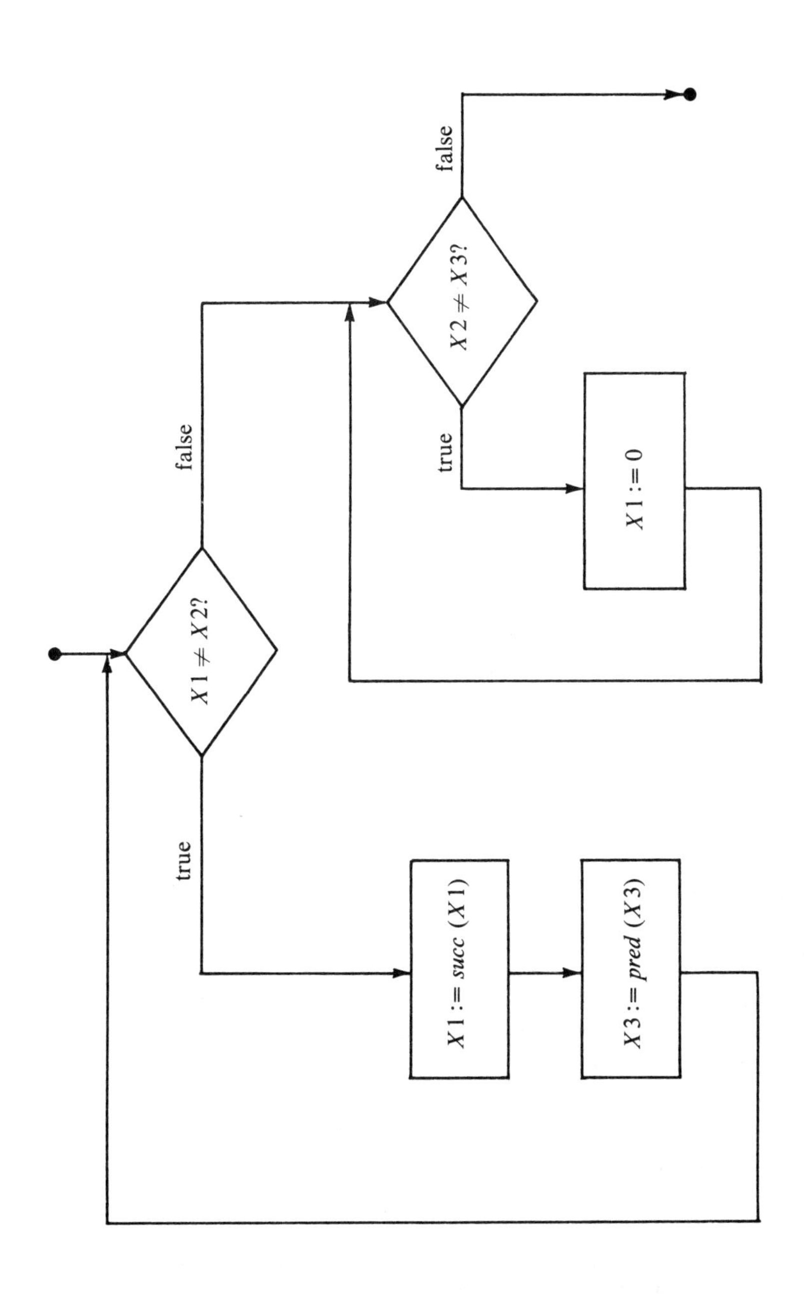
X 1 ≠ X 2?
true
false
X 2 ≠ X 3?
true
false
X 1 := 0
X 1 := succ (X 1)
X 3 := pred (X 3)

OPERATIONAL SEMANTICS OF **while**-PROGRAMS

Given a k-variable program P, we wish to interpret P as an agent for computing a j-ary function

$$\varphi_P : \mathbf{N}^j \rightarrow \mathbf{N}.$$

(Our convention throughout the book is to denote the partial function computed by a program P by φ_P.) We accomplish this using the machinery of the previous subsection. Suppose that the variables in P are (a subset of) $X1, \ldots, Xk$.

6 Definition. The j-ary semantics function for a k-variable P, $\varphi_P : \mathbf{N}^j \rightarrow \mathbf{N}$ is defined as follows. Given input vector $(a_1, \ldots, a_j)$, $\varphi_P(a_1, \ldots, a_j)$ is evaluated according to the following rules, with two cases arising.

Case 1. Suppose $k \geqslant j$. Then $\varphi_P(a_1, \ldots, a_j)$ is evaluated by applying P to the initial state vector

$$(a_1, \ldots, a_j, \underbrace{0, \ldots, 0}_{(k-j)\ 0\text{'s}})$$

If and when P halts on this state vector, the value of $\varphi_P(a_1, \ldots, a_j) = b$, where b is the value, upon termination, of $X1$.

Case 2. $k < j$. To compute $\varphi_P(a_1, \ldots, a_j)$ in this case, apply P to the initial state vector

$$(a_1, \ldots, a_k)$$

ignoring the last $(j-k)$ arguments. Again, if and when P halts on this vector, the value of $\varphi_P(a_1, \ldots, a_j) = b$, where b is the final value of $X1$.

In either case, if P fails to halt on the initial state vector, then the value of $\varphi_P(a_1, \ldots, a_j)$ is undefined, and we write in this case $\varphi_P(a_1, \ldots, a_j) = \perp$ ($\perp$ is the symbol for the "undefined value")[2].

Notice that according to this definition **while**-programs with no variables, i.e., **begin end**, **begin begin end end**, etc., have the property that for a program P in this class,

$$\varphi_P(a_1, \ldots, a_j) = a_1$$

no matter what the choice of j. (Recall that we require $k > 0$.)

[2] In this book we use two symbols, $\uparrow$ and $\perp$, in relation to diverging computations. Both are used in the literature, and we have not tried to disqualify one in favor of the other. If there is a distinction to be made between the two, it is in the following sense: $\uparrow$ indicates that a computation diverges, while $\perp$ denotes the *outcome* of a diverging computation. Thus if φ is a computable function, we may write $\varphi(a)\uparrow$ or $\varphi(a) = \perp$; the statements are synonymous.

7 Definition. We say that a function

$$\psi : \mathbf{N}^j \to \mathbf{N}$$

is *effectively computable* (or more succinctly, *computable*) if $\psi = \varphi_P$ for some **while**-program P. If we want to stress the number of input variables, we write $\varphi_P^{(j)} : \mathbf{N}^j \to \mathbf{N}$.

8 Example. Consider the following program P:

```
begin
    while X1 ≠ X2 + X3 do
        begin X1 := succ(X1);
              X2 := pred(X2)
        end
end
```

On input $(3,2,1)$, the value of $\varphi_P^{(3)} : \mathbf{N}^3 \to \mathbf{N}$ is 3; on input $(1,2,3)$, the value of $\varphi_P^{(3)}$ is 3 again; and on input $(1,1,1)$, the value of $\varphi_P^{(3)}$ is $\perp$ (undefined).

On the other hand, for $\varphi_P^{(2)} : \mathbf{N}^2 \to \mathbf{N}$, on all input pairs $X3$ is set to 0 and hence plays no role in the computation. In this case $\varphi_P^{(2)}(1,2) = \perp$, and $\varphi_P^{(2)}(1,3) = 2$.

It is of course easy to extend the definition of computable function to include functions of the form $\varphi_P : \mathbf{N}^j \to \mathbf{N}^m$ where P is a k-variable program with $k \geqslant m$, and where $\varphi_P(a_1, \ldots, a_j)$ is read off from the final values of $X1, \ldots, Xm$ in a computation with input vector determined as in Definition 6. However, for ease of exposition, we shall usually consider only one-dimensional output, $m = 1$. The reader may find it a useful exercise to develop the m-dimensional version of the results that follow.

Summary of the semantics of while-programs

Assume all variables in **while**-program P are from the set $\{X1, X2, \ldots, Xk\}$ for some $k \geqslant 1$. For every $j \geqslant 1$, the j-ary function $\varphi_P : \mathbf{N}^j \to \mathbf{N}$ computed by P is defined according to two cases.

- Case 1: $k \geqslant j$.

If and when P halts on input vector

$$(a_1, \ldots, a_j, \underbrace{0, \ldots, 0}_{(k-j)})$$

the value of $\varphi_P(a_1, \ldots, a_j)$ is b, where b is the value assigned to $X1$ upon termination.

- Case 2: $k < j$.

If and when P halts on input vector $(a_1, \ldots, a_k)$, the value of $\varphi_P(a_1, \ldots, a_j)$ is b, where b is the value assigned to $X1$ upon termination.

We close this section by rephrasing Church's Thesis in the context of our programming approach to computability. The completeness of the specification of algorithms via **while**-programs, as explained in Section 1.1, is equivalent to the following formulation of Church's Thesis.

9 Church's Thesis Revisited. *Any partial function from the natural numbers to the natural numbers that can in one way or another be algorithmically defined is also computed by some* **while**-*program.*

Church's Thesis has the following crucial consequence for us: If we can establish that a number-theoretic function θ is algorithmic in nature, then θ is in fact computable by some **while**-program; and we may use this fact freely, even if we do not want to exhibit the indicated **while**-program explicitly. This is valuable because it will often be sufficient to know that a function θ is computable without knowing an actual **while**-program that computes it. In other words, we will often use Church's Thesis to give a *proof outline* rather than a proof. In every case where we simply sketch why some θ is algorithmic, the reader may rest assured (or take it as a challenge to prove) that a full proof can be given that demonstrates that a **while**-program for θ does indeed exist.

Exercises for Section 2.3

1. Show that no **while**-program with exactly one variable can compute the function $f(x) = 2 * x$. (Note: Do not use macros.)
2. We use the definition of "straight-line programs" given in Exercise 1 of Section 2.1. Classify all functions from **N** to **N** computed by:
   (a) Straight-line programs with exactly one variable.
   (b) Straight-line programs in general, with no restriction on the number of variables they use.
3. For this exercise, review the material of Section 1.2.
   (a) Show that it is decidable whether an arbitrary straight-line program halts on an arbitrary input. (See the definition of "straight-line programs" in the preceding exercise.)
   (b) Show that it is decidable whether an arbitrary **while**-program with exactly one variable halts on an arbitrary input. (*Hint*: See Exercise 1 of Section 2.1.)
4. Show that any total function $f: \mathbf{N} \to \mathbf{N}$ which is zero "almost everywhere", i.e., zero for all but finitely many values of its argument, is effectively computable.
5. Show that the function $f: \mathbf{N} \times \mathbf{N} \to \mathbf{N}$ is computable by a **while**-program, where

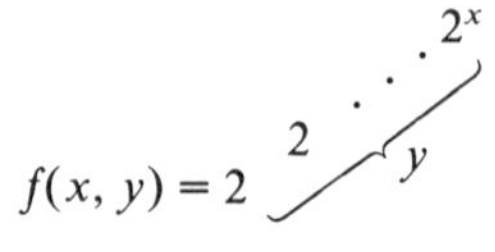

$$f(x, y) = 2^{2^{\cdot^{\cdot^{\cdot^{2^x}}}}} \Big\} y$$

i.e., the value of $f(x, y)$ is "a stack of y exponentiated 2's topped off with an x".

6. (a) Show that the square root function $\sqrt{x}$ over the natural numbers is effectively computable. (This $\sqrt{x}$ is integer valued, e.g., $\sqrt{4} = \sqrt{5} = 2$ and $\sqrt{10} = 3$.)
   (b) The *excess-over-a-square* function defined by $excess(x) = x \dot{-} [\sqrt{x}\,]^2$. Show that for every $y \in \mathbf{N}$ there are infinitely many $x \in \mathbf{N}$ for which $excess(x) = y$. Write a **while**-program that computes the function *excess*.

7. Show that the divisibility predicate is effectively computable, that is, there is a **while**-program that computes the following function:
$$x \mid y = \begin{cases} 1, & \text{if } x \text{ divides } y \text{ exactly;} \\ 0, & \text{otherwise.} \end{cases}$$

8. Show that each of the following functions is computable by a **while**-program:
   (a) $pr(x)$ = "the xth prime number", given that 2 is the 0th prime number.
   (b) $prime(x) = \begin{cases} 1, & \text{if } x \text{ is a prime number;} \\ 0, & \text{if } x \text{ is not a prime number.} \end{cases}$
   (c) $exp(x, y)$ = "the exponent of the prime number $pr(x)$ in the prime decomposition of y", with the agreement that $exp(x, 0) = 0$ for all $x \in \mathbf{N}$.

9. For this exercise we define *composition of functions* as follows (which is slightly different from the definition of Section 1.1). Let $\psi : \mathbf{N}^k \to \mathbf{N}$ be an arbitrary k-ary function and $\zeta_1, \ldots, \zeta_k$ be arbitrary functions each from $\mathbf{N}^l$ to $\mathbf{N}$. The composition of ψ with $\zeta_1, \ldots, \zeta_k$—written $\psi(\zeta_1, \ldots, \zeta_k)$—is an l-ary function defined by:
$$\psi(\zeta_1, \ldots, \zeta_k)(x_1, \ldots, x_l) = \psi(\zeta_1(x_1, \ldots, x_l), \ldots, \zeta_k(x_1, \ldots, x_l)).$$
The operation of *minimization* is defined as follows. Given an arbitrary total function $f : \mathbf{N}^{k+1} \to \mathbf{N}$, we define a k-ary function $\theta : \mathbf{N}^k \to \mathbf{N}$ by means of *minimization* if
$$\theta(x_1, \ldots, x_k) = \begin{cases} y, & \text{if } y \text{ is the smallest number such that } f(x_1, \ldots, x_k, y) = 1; \\ \perp, & \text{otherwise;} \end{cases}$$
and we write $\theta(x_1, \ldots, x_k) = \mu y\, [f(x_1, \ldots, x_k, y) = 1]$. Note that even though f is total, θ may not be. This cannot happen, however, if θ [now a $(k+1)$-ary function] is defined instead by *bounded minimization*:
$$\theta(x_1, \ldots, x_k, z) = \begin{cases} y, & \text{if } y \text{ is the smallest number} \leqslant z \text{ such that } f(x_1, \ldots, x_k, y) = 1; \\ z, & \text{if no such } y \text{ exists.} \end{cases}$$
And we now write $\theta(x_1, \ldots, x_k, z) = \mu y \leqslant z\, [f(x_1, \ldots, x_k, y) = 1]$.
   Prove that the class of effectively computable functions is closed under the operations of composition, minimization, and bounded minimization.

10. We use the definition of "minimization" in the previous exercise, with the total function f replaced by an arbitrary function ζ (zeta). Show that there is a function $\zeta : \mathbf{N}^2 \to \mathbf{N}$ which is not total, but $\theta(x) = \mu y\, [\zeta(x, y) = 1]$ is total. (One can show that the computable functions are not closed under this modified minimization [Rogers (1967), Exercise 2-13].)

11. Given an arbitrary function $\zeta : \mathbf{N} \to \mathbf{N}$, define the unary functions $A\zeta$, $B\zeta$, and $C\zeta$ as follows:

$$A\zeta(x) = \begin{cases} y, & \text{if } y \text{ is the smallest number such that } \zeta(y) = x; \\ 1, & \text{if no such } y \text{ exists.} \end{cases}$$

$$B\zeta(x) = \begin{cases} 0 & \text{if } \zeta(x) = 0; \\ 1, & \text{if } \zeta(x) \geqslant 1 \text{ and defined;} \\ \bot, & \text{if } \zeta(x) = \bot. \end{cases}$$

$$C\zeta(x) = \begin{cases} y, & \text{if } \zeta(x) \text{ is defined and } y \text{ is the length of } \zeta(x) \text{ written in binary;} \\ \bot, & \text{if } \zeta(x) = \bot. \end{cases}$$

(a) Show that the class of computable unary functions over **N** is closed under the operations B and C.
(b) Show that the class of computable unary functions, which are further restricted to have infinite ranges and be strictly increasing, is closed under operation A but not operations B and C.
[*Hint for* (b): Use Church's Thesis liberally.]

12. (a) Exhibit a bijection from **N** to **N** which is effectively computable. Prove also that there are bijections from **N** to **N** which cannot be effectively computed.
(b) Show that the class of computable bijections from **N** to **N** is closed under composition and taking inverses. (If $f : \mathbf{N} \to \mathbf{N}$ is a bijection then the inverse of f is the $g : \mathbf{N} \to \mathbf{N}$ such that $g \circ f$ is the identity function.)

13. Let the function $A : \mathbf{N} \times \mathbf{N} \to \mathbf{N}$ be defined recursively by

$$A(0, y) = y + 1,$$
$$A(x + 1, 0) = A(x, 1),$$
$$A(x + 1, y + 1) = A(x, A(x + 1, y)).$$

The function A is known as *Ackermann's function*.

(a) Prove that Ackermann's function is total.
(b) Write a **while**-program that computes Ackermann's function.
(c) Prove

$$A(1, y) = y + 2, \quad A(2, y) = 2y + 3, \quad A(3, y) = 2^{y+3} - 3,$$

$$A(4, y) = \left.2^{2^{\cdot^{\cdot^{\cdot^{2}}}}}\right\}{\scriptstyle y+3} - 3.$$

In each case, determine the "time" required (or at least a tight lower bound for it) to compute the value, no matter what **while**-program is used to compute the function A. "Time" is measured by the number of basic instructions carried out, each basic instruction being of the form $X := 0$ or $X := succ(Y)$ or $X := pred(Y)$ or $X \neq Y$?

CHAPTER 3

Enumeration and Universality of the Computable Functions

We began our discussion of the Halting Problem in Chapter 1 by showing how to make a systematic listing, based on ASCII character codes, of all PASCAL programs. Each PASCAL program had a unique number, or index. This number was obtained by concatenating the ASCII bit strings representing the resulting binary sequence as an integer in binary. Of course, not every number corresponded to a legal program, or even to a legal sequence of ASCII blocks. Under these circumstances an invalid program in our listing was interpreted, by default, to be a particular program for the empty function. In Section 3.1 we follow a similar plan and present an enumeration of the **while**-programs. Then, in Section 3.2, we develop an "interpreter", also called a universal program, which can decode the index n of a program $\mathbf{P}_n$ to process data just as $\mathbf{P}_n$ would have done. Sections 3.3 and 3.4 develop "subroutines" for the construction of the interpreter showing, respectively, how string-processing functions may be simulated by numerical functions and how vectors of natural numbers may be coded by a single such number.

3.1 The Effective Enumeration of **while**-Programs

Since we will only be interested in character strings that represent programs, a six-bit binary code is sufficient to represent the character set of the language of **while**-programs. We give the code in Table 1.

Table 1

Symbol	Binary Code	Decimal Code
begin	100000	32
end	100001	33
while	100010	34
do	100011	35
:=	100100	36
;	100101	37
succ	100110	38
pred	100111	39
≠	101000	40
(	101001	41
)	101010	42
X	101011	43
0	101100	44
1	101101	45
2	101110	46
3	101111	47
4	110000	48
5	110001	49
6	110010	50
7	110011	51
8	110100	52
9	110101	53

In Table 1 there is a unique code assigned to every symbol (or invariant sequence of symbols) in the basic character set of our PASCAL fragment. The code for a variable name such as $X13$ is built up from the codes for "X", "1", and "3". Other variable names—X, $Y1$, $Y2$, Z, TEMP1, etc.—should be regarded as abbreviations for variables of the form Xn, where n is a non-negative integer in decimal notation.

Our standard listing of **while**-programs

$$\mathbf{P}_0, \mathbf{P}_1, \mathbf{P}_2, \ldots, \mathbf{P}_n, \ldots$$

is constructed as follows: $\mathbf{P}_n$ is the program obtained by converting integer n to binary, dividing this representation of n into six-bit blocks, and interpreting each block as a symbol or sequence of symbols in the syntax of **while**-programs. We will use boldface $\mathbf{P}$ only in conjunction with our standard listing of **while**-programs, leaving P, Q, and R (sometimes subscripted) as variables ranging over the set $\{\mathbf{P}_0, \mathbf{P}_1, \mathbf{P}_2, \ldots\}$. The expressions "program n", "program with index n", and $\mathbf{P}_n$ will be synonymous.

Of course not all non-negative integers n lead to legitimate **while**-

programs. $\mathbf{P}_{2082}$, for example, corresponds to the string

100000	100010
begin	**while**

Likewise, since the binary code of $\mathbf{P}_{294}$ is 100100110 whose length is not a multiple of 6, $\mathbf{P}_{294}$ is not a bona-fide **while**-program. Nevertheless, we do not exclude these strings from the listing, but instead simply classify them as *syntactically incorrect* **while**-programs. (Recall that a syntactically correct **while**-program is one which is written according to the rules of Section 2.1.) Following our method of Section 1.2, let us agree that all non-programs of this sort are to be replaced by a particular program for the empty function.

Our coding scheme also gives us a simple way to find a program's *index*, i.e., its position in our listing of all **while**-programs. For example, the program

begin
$X1 := succ(X1)$
end

has as its binary code:

100000	101011	101101	100100	100110	101001	101011	101101	101010	100001
begin	X	1	:=	*succ*	(	X	1	)	**end**

The index of the above program is the integer denoted by the corresponding binary code. It falls somewhere between 2^{59} and 2^{60}. Fortunately the development to follow is not affected by the magnitude of program indices —all that really matters is that different strings have different numbers.

We now have a systematic way of assigning an index to every **while**-program, and given any index we can also retrieve the unique **while**-program which it identifies.

The process of systematically assigning natural numbers to syntactic objects is sometimes called an *arithmetization of syntax* or a *Gödel numbering*. This idea has played an extremely important role in the foundations of mathematics. We shall discuss another aspect of this idea in Section 7.3, where we consider a classical result of mathematical logic, namely, Gödel's Incompleteness Theorem.

We now say a few words about what it means for a set to be "effectively enumerated." This notion is best illustrated by the process of writing the natural numbers in increasing order of magnitude. Since we can write 0 and we can also add 1 to any integer (in binary or decimal notation, for example), we can enumerate in sequence all the natural numbers:

$$0, 1, 2, \ldots, n, n + 1, \ldots .$$

This enumeration is called *effective* because it can be carried out in a stepwise mechanistic manner, using the constant function 0 and the function *succ* only, both assumed to be effective or algorithmic in the context of our theory. There are sets other then **N** that can also be effectively enumerated, and we shall study them in detail in Chapter 7, where they are called "recursively enumerable sets".

The arithmetization of a countably infinite set of syntactic objects (in particular, the set of all **while**-programs) tells us that this set can be effectively enumerated—hence the title of this section. Indeed, as we effectively enumerate all natural numbers, we can effectively enumerate all **while**-programs at the same time according to the translation $n \mapsto \mathbf{P}_n$ defined above:

$$0, \ 1, \ 2, \ \ldots, n, \ \ n+1, \ldots,$$
$$\mathbf{P}_0, \mathbf{P}_1, \mathbf{P}_2, \ldots, \mathbf{P}_n, \mathbf{P}_{n+1}, \ldots .$$

This is not the only effective enumeration of all **while**-programs; for every arithmetization of **while**-programs there is a corresponding effective enumeration (Exercises 1 and 2).

Enumerating the computable functions

Now that we have settled on a fixed (effective) enumeration of all **while**-programs, we can give a similar listing of all computable functions from $\mathbf{N}^j$ to $\mathbf{N}$. Recall, from Definition 6 in Section 2.3, that we may associate a function $\varphi_P^{(j)}: \mathbf{N}^j \to \mathbf{N}$ with each **while**-program P—namely, if P is a k-variable program, load $(a_1, \ldots, a_j)$ into $X1, \ldots, Xj$ (discarding $a_{k+1}, \ldots, a_j$ if $j > k$; setting $Xj+1, \ldots, Xk$ to 0 if $k > j$), run P, then if and when the computation terminates, read $\varphi_P^{(j)}(a_1, \ldots, a_j)$ from $X1$. Thus, our enumeration $\mathbf{P}_0, \mathbf{P}_1, \ldots, \mathbf{P}_n, \ldots$ of all **while**-programs yields an enumeration

$$\varphi_0^{(j)}, \varphi_1^{(j)}, \ldots, \varphi_n^{(j)}, \ldots,$$

where $\varphi_n^{(j)}$ is the j-ary function computed by program $\mathbf{P}_n$. This will be our standard enumeration throughout the book. The one-variable computable functions

$$\varphi_0^{(1)}, \varphi_1^{(1)}, \ldots, \varphi_n^{(1)}, \ldots$$

play a particularly important role in computability theory, since any j-ary partial function can be viewed as a one-variable function by appropriate coding and decoding of the j-ary input vectors, as we shall study in Section 3.4. Hence when we write

$$\varphi_0, \varphi_1, \ldots, \varphi_n, \ldots,$$

we will mean the computable one-variable functions, unless we state otherwise.

If $\psi: \mathbf{N}^j \to \mathbf{N}$ is computable, say $\psi = \varphi_i^{(j)}$, we shall say that i is an *index* for ψ.

Let us have a first look at our list of the computable functions. Since the list is indexed by the natural numbers, there are evidently only countably many computable functions mapping $\mathbf{N}^j$ to $\mathbf{N}$, in contrast to the uncountable number of functions of any sort from $\mathbf{N}^j$ to $\mathbf{N}$.

Moreover if ψ is computable, say $\psi = \varphi_i^{(j)}$, then ψ must in fact have infinitely many indices. This must happen because we can always pick one program P for ψ, and produce an infinite sequence of modified programs

$$Q_1, Q_2, \ldots, Q_n, \ldots$$

all of which are syntactically different, and all of which do just what P does. Let Xu be a variable in P. Then to construct Q_n, insert, just after the first **begin** of P, some $2n$ instructions, of which the first n are $Xu := succ(Xu)$, and the last n are $Xu := pred(Xu)$.

In particular, there are infinitely many natural numbers e for which φ_e is the empty function, and also infinitely many e for which φ_e is the constant function identically equal to 1.

The main result of this chapter is that there is also an index u for which

$$\varphi_u^{(2)}(x, y) = \varphi_x(y),$$

for all possible values of x and y. This u encodes a *universal* function—a function whose behavior simulates the behavior of any one-variable function on any input. To establish the existence of this function we shall actually construct a **while**-program which is an *interpreter* for **while**-programs. Computer scientists should not find such a construction paradoxical. After all, interpreters and compilers in everyday computer science are often written in the language they are intended to operate on.

The Halting Program Revisited

Let us reexamine the most fundamental result of our subject, the unsolvability of the Halting Problem, using the machinery developed so far.

1 Theorem (The Unsolvability of the Halting Problem). *There is no total computable function f such that*

$$f(i) = \begin{cases} 1, & \text{if } \varphi_i(i) \text{ converges;} \\ 0, & \text{if } \varphi_i(i) \text{ fails to converge.} \end{cases}$$

PROOF. Suppose f were computable Consider the function $\psi : \mathbf{N} \to \mathbf{N}$ (the function computed by program *confuse* of Section 1.2).

$$\psi(i) = \begin{cases} \perp, & \text{if } f(i) = 1; \\ 1, & \text{if } f(i) = 0. \end{cases}$$

If there is a **while**-program *halt* to compute f, then the following is a **while**-program to compute ψ:

begin while *halt* $(X1) = 1$ **do** $X1 := X1$; $X1 := 1$ **end**

Let this program have index e. But then $\psi(i)$ must be $\varphi_e(i)$ and so $\psi(e)$, i.e., $\varphi_e(e)$ must have contradictory behavior. That is, if $\psi(e)$ halts, then it diverges, but if it fails to halt, then it converges. We conclude that f cannot be computable. □

We shall explore the implications of the unsolvability of this problem in considerable detail in later chapters.

Exercises for Section 3.1

1. Using an enumeration of natural numbers based on their representation in base 22, devise an alternative arithmetization of **while**-programs. (Note that the basic character set of **while**-programs contains 22 symbols, as shown in Table 1.)

2. The arithmetization of **while**-programs given in the text, as well as that in the preceding exercise, depended on our explicitly choosing a system of notation for the natural numbers. The arithmetization given in the present exercise only uses the prime decomposition of natural numbers, and does not depend on whether natural numbers are written in decimal, binary, base 22, or any other notation.

   If $\mathbb{S}$ is a statement in the language of **while**-programs, let us denote by $[\mathbb{S}]$ the index assigned to $\mathbb{S}$. Let us assign indices according to the following inductive definition:

   (1) [**begin end**] $= 1$.
   (2) For all, $i, j \geqslant 1$:
   $$[Xi := 0] = 2^i, [Xi := succ(Xj)] = 3^i 5^j, [Xi := pred(Xj)] = 7^i 11^j.$$
   (3) For all $i, j \geqslant 1$ and statements $\mathbb{S}$:
   $$[\textbf{while } Xi \neq Xj \textbf{ do } \mathbb{S}] = 13^i 17^j 19^{[\mathbb{S}]}$$
   (4) For all statements $\mathbb{S}_1, \ldots, \mathbb{S}_m$ and $m \geqslant 1$:
   $$[\textbf{begin } \mathbb{S}_1 ; \ldots ; \mathbb{S}_m \textbf{ end}] = pr(8)^{[\mathbb{S}_1]} pr(9)^{[\mathbb{S}_2]} \ldots pr(7+m)^{[\mathbb{S}_m]},$$

   where $pr(8) = 23$ and, in general, $pr(i)$ is the ith prime number (see Exercise 2.3.8).

   (a) Show that the function [ ]: {**while**-programs} $\rightarrow \mathbf{N}$ is one-to-one. Is it onto?
   (b) Outline the steps of a procedure which, given an arbitrary $n \in \mathbf{N}$, will decide whether n is the index of a **while**-program.

3. Deduce from the unsolvability of the Halting Problem that the function f defined as follows is not effectively computable:
   $$f(i) = \begin{cases} 1, & \text{if } \varphi_i(0) \text{ converges;} \\ 0, & \text{if } \varphi_i(0) \text{ fails to converge.} \end{cases}$$
   (*Hint*: Show first that for every **while**-program P, we can construct a **while**-program P' such that P converges on its own index as input if and only if P' converges on input zero.)

4. Let $\varphi_0, \varphi_1, \ldots, \varphi_n, \ldots$ be our standard enumeration of the unary computable functions. Define a function $\theta : \mathbf{N} \to \mathbf{N}$ as follows.

$$\theta(n) = \begin{cases} \sum_{i \leq n} \varphi_i(n), & \text{if each of } \varphi_0(n), \ldots, \varphi_n(n), \text{ is defined;} \\ \perp, & \text{otherwise.} \end{cases}$$

   Prove that θ is effectively computable.

5. Consider the function

$$f(i) = \begin{cases} 1, & \text{if } \varphi_i(i) = 1; \\ 0, & \text{otherwise.} \end{cases}$$

   Is f effectively computable? Justify your answer.

3.2 Universal Functions and Interpreters

Consider our standard enumeration of the j-ary computable functions

$$\varphi_0^{(j)}, \varphi_1^{(j)}, \ldots, \varphi_m^{(j)}, \ldots .$$

This section is devoted to the proof of:

1 Theorem (The Enumeration Theorem). *For every $j \geq 1$, there exists a computable function $\Phi : \mathbf{N}^{j+1} \to \mathbf{N}$ which is* universal *for the given enumeration of the j-ary functions in that for all $e \in \mathbf{N}$ and $(a_1, \ldots, a_j) \in \mathbf{N}^j$,*

$$\Phi(e, a_1, \ldots, a_j) = \varphi_e(a_1, \ldots, a_j).$$

As the first argument of Φ is made to vary over $\mathbf{N}$, Φ enumerates all the j-ary computable functions—hence the name "enumeration theorem". This result is proved by constructing a **while**-program $\mathbf{P}_u$ (u for universal) which is an interpreter for **while**-programs: Given the index e encoding the program $\mathbf{P}_e$ and the data vector $(a_1, \ldots, a_j)$, $\mathbf{P}_u$ will simulate $\mathbf{P}_e$ to process the data, yielding the result $\varphi_e(a_1, \ldots, a_j)$ (and never terminating if $\varphi_e(a_1, \ldots, a_j)$ is itself undefined). We shall sketch the proof in considerable detail, even though the existence of interpreters for programming languages should make this result seem highly plausible.

Before outlining the construction of a **while**-program which computes Φ, we use Church's Thesis to motivate the Enumeration Theorem. An algorithm to compute the universal function $\Phi : \mathbf{N}^{j+1} \to \mathbf{N}$ is very simply described:

1. Given natural numbers $e, a_1, \ldots, a_j$, write the **while**-program that e encodes, i.e., $\mathbf{P}_e$.

2. Carry out the computation of $\mathbf{P}_e$ on input vector $(a_1, \ldots, a_j)$.
3. If and when this computation terminates, return the value in variable $X1$ of $\mathbf{P}_e$, which is by definition the value of $\varphi_e^{(j)}(a_1, \ldots, a_j)$.

Each of these three steps can be executed in a systematic and deterministic way. Since this algorithm defines Φ, Church's Thesis asserts that Φ is effectively computable by some **while**-program.

Here of course we have bypassed all the programming details of the universal **while**-program that computes Φ. Since the Enumeration Theorem is the foundation for much of the subsequent development of this book, however, we devote the rest of this section to a demonstration that we can rewrite this well-specified algorithm in the language of **while**-programs.

For the remainder of this section, we describe certain macros and show how they can be combined to yield a **while**-program interpreter. In the rest of the chapter we develop **while**-programs for string manipulation (Section 3.3) and for encoding several integers into a single integer and decoding the results (Section 3.4). Making use of these **while**-programs, we can outline, in considerable detail, the macros that we need to write our interpreter.

Our interpreter will take $(e, a_1, \ldots, a_j)$ as input and return $\varphi_e(a_1, \ldots, a_j)$ if this result is defined, and never terminate otherwise. As a first step, then, we will have our interpreter check whether e is the index of a syntactically well-formed **while**-program. If e is not such an index, then $\mathbf{P}_e$ defines the empty function, and our interpreter simply goes into an infinite loop, never returning a result. If e does define a genuine **while**-program, then, as we shall spell out below, we must translate $\mathbf{P}_e$ into a form that the interpreter can use. We thus need to establish the following:

2 Proposition. *There is an effectively computable function legal*: $\mathbf{N} \to \{0, 1\}$ *such that*

$$legal(e) = \begin{cases} 1, & \text{if } e \text{ encodes a syntactically correct } \textbf{while}\text{-program}; \\ 0, & \text{otherwise.} \end{cases}$$

The proof of this proposition is delayed to Section 3.3.

Suppose, now, that $\mathbf{P}_e$ is syntactically correct. Then, in particular, the text of $\mathbf{P}_e$ uses certain variables Xj. Let us say that l is the largest number for which Xl appears in $\mathbf{P}_e$. Now l must be finite, but since e is arbitrary, l may be arbitrarily large. In particular, l may be far larger than the largest j for an Xj of our interpreter. In other words, our interpreter *cannot* interpret $\mathbf{P}_e$ by using the variable Xi wherever $\mathbf{P}_e$ uses that Xi, because the interpreter is a fixed program with a fixed stock of variables. Thus the first step in coding $\mathbf{P}_e$ into an intepretable form is to see how an arbitrary number of variables can be coded into a fixed set of variables. We shall do this in Section 3.4, where we shall prove the following result.

3 Proposition. *For all $j \geqslant 1$, there is a natural number r and a total computable function short*: $\mathbf{N} \to \mathbf{N}$ (*which depends on j*) *such that*

(a) $\varphi_e^{(j)} = \varphi_{short(e)}^{(j)}$.
(b) Program $\mathbf{P}_{short(e)}$ uses $j + r$ variables.

If we think about the operations discussed above, we see that we have a crucial need for macros for *string-processing*. It is clear that, given an e which encodes a program, we want to process it not so much as a number, but as a string, reading off six bits at a time to process them as a single symbol according to Table 1 of Section 3.1. For example, to find the largest l for which Xl occurs in the text of e, we must be able to scan e, six bits at a time from left to right, until we find the code for an X; then we must continue six bits at a time so long as we meet six-bit codes for digits. The resultant encoding of a string of digits is our first candidate for l. And so the process will continue, as we find new such strings, retain the largest, until we reach the end of the binary string that is e. We shall explicitly study macros for string processing in Section 3.3. In the rest of this section, providing an overview of the interpreter, we shall freely use the language of string-processing.

Returning to the design of the interpreter, then, we have stated (Proposition 3) that it can translate $\mathbf{P}_e$ into $\mathbf{P}_{short(e)}$—call it P for short—which has a restricted set of variables. However, interpreting P is still not straightforward. Consider the semicolon indicated by the arrow in the text below:

$$\textbf{begin while } X1 \neq X2 \textbf{ do while } X1 \neq X4 \textbf{ do } X1 := succ(X1); X2 := X4 \textbf{ end}$$

$$\uparrow$$

There is nothing in the *local* text to indicate whether to go back to an earlier test, or to continue to the next instruction. Nor is it easy to tell just where in the text to return to, if the next instruction is not appropriate. To solve this problem, the interpreter translates P into *quadruple form*, in which each instruction is given an address both for itself and for the instruction to be executed next. This use of addressed instructions moves us from ALGOL-like languages (such as our fragment of PASCAL) to assembly-like languages. We now examine this transformation on a specific example.

Consider the **while**-program Q whose flow diagram is shown on page 54, where we also attach a unique label to each instruction as well as to the exit point. In this labelled flow diagram, each assignment instruction has a uniquely labelled successor. The same is true of the label of a test instruction, provided a Boolean value (*true* or *false*) is also specified. This allows us to rewrite Q in the form of a finite sequence of quadruples of the form

(label; instruction; false label; true label).

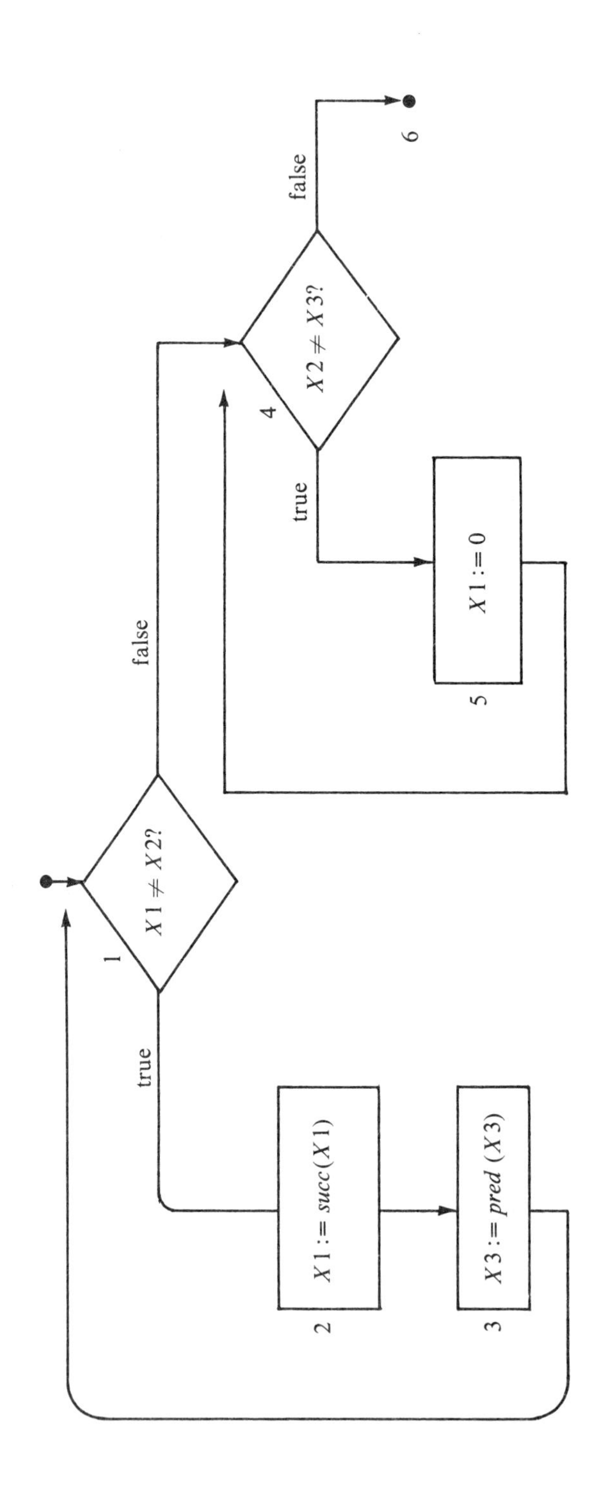
1
X1 ≠ X2?
true
false
2
X1 := succ(X1)
3
X3 := pred (X3)
4
X2 ≠ X3?
true
false
5
X1 := 0
6

The following sequence of quadruples uniquely determines **while**-program Q:

$$(1;\ X1 \neq X2;\ 4;\ 2)$$
$$(2;\ X1 := succ(X1);\ 3;\ 3)$$
$$(3;\ X3 := pred(X3);\ 1;\ 1)$$
$$(4;\ X2 \neq X3;\ 6;\ 5)$$
$$(5;\ X1 := 0;\ 4;\ 4)$$

In the above quadruples whose first entries are 2, 3, and 5 (labels of assignment instructions in Q), we have set the third and fourth entries to the same value. Note that the concatenation of all five quadruples is a finite string over our basic character set (Table 1 of Section 3.1), which should be derivable from the text of Q using our string-processing machinery.

We can apply the preceding transformation to any **while**-program P in general—but now we consider P in the form of a text and not a flow diagram. Our convention for labelling instructions (implicit in our treatment of Q) is to attach 1 to the first instruction of P, and then successively attach the next label to the next encountered instruction in reading the text of P top-down and left-to-right. In this fashion, if n is the number of instructions in P, label $n + 1$ identifies the exit of P. We also agree to call the quadruple as defined above a *quadruple* (naturally enough!).[1]

4 Proposition. *There is a total computable function quad*: $\mathbf{N} \to \mathbf{N}$ *such that if e encodes a syntactically correct* **while**-*program P, then quad(e) encodes the sequence of quadruples of P.*

Proof. We define the string-processing version of *quad* in seven stages:

(A) Scan the text of P from left-to-right, replacing every test instruction

[1] Some care should be applied in differentiating labels from their numerical codes. Quadruples as defined here are strings of symbols over our basic character set (Table 1). Hence if we attach label 8, say, to some instruction, we in fact attach to it number '8' = 52 (the numerical code of the symbol 8). The next available label is 9, whose numerical code is 53 according to Table 1, and in this case we can go from the numerical code of 8 to that of 9 by one application of the successor function, i.e., '9' = *succ*('8'). What about the next label 10? Its numerical code '10' = $45 \times 2^6 + 44 = 2924$, according to Table 1, is not equal to *succ*('9') = 54. We encounter a similar problem when we go from label 19 to label 20, or from 29 to 30, or from 99 to 100, etc. Hence, when we want to generate the next available label (in fact, its numerical code) using the number-theoretic functions at our disposal, we should remember that it is not always possible by application of *succ* only once.

There are several ways of amending our instruction-labelling process in order to avoid the complication mentioned above. The reader may wish to make such amendments in order to simplify some of the programming details in Proposition 4. For example, instead of labelling instructions with the sequence $1, 2, 3, \ldots,$ he or she may use labels from the sequence $1, 11, 111, \ldots,$ whose corresponding numerical codes are easily generated as: 45, $45 \times 2^6 + 45$, $45 \times 2^{12} + 45 \times 2^6 + 45, \ldots$.

$Xi \neq Xj$ by the string

$$(; Xi \neq Xj; ;)$$

and every assignment instruction $Xi := g(Xj)$, where $g \in \{succ, pred, 0\}$, by the string

$$(; Xi := g(Xj); ;)$$

The outcome of stage A is a string P_1 where every instruction of P has been changed to a quadruple whose first, third, and fourth entries are not yet specified.

(B) Scan the text of P_1 from left-to-right, inserting the appropriate label whenever the pair (; is encountered. The combination of left-hand parenthesis immediately followed by a semicolon indicates that the first entry of a quadruple is missing. We start with label 1 and increment by 1 every time a new label is to be inserted. Call the outcome of this stage P_2.

The remaining stages insert the appropriate third and fourth entries (the "false label" and the "true label") in each of the quadruples. To simplify this process in stages D, E, and F, Stage C eliminates redundant **begin-end** pairs[2] as well as empty compound statements.

(C) Scan the text of P_2 from left-to-right, enclosing the body of every **while** statement in a **begin–end** pair (if it is not already enclosed in such a pair). Scan the text of P_2 again, replacing every empty compound statement of the form "**begin end**" by "**begin** $X1 := succ(X1)$; $X1 := pred(X1)$ **end**". Scan the text of P_2 once more, looking for every **begin** not immediately preceded by **do**—eliminate every such **begin** and the matching **end**. Call the outcome of this stage P_3.

Note that every **begin** in P_3 is now the entry point of the body of a **while** statement (which is never empty in P_3). Hence every **end** in P_3 is also the exit point of the body of a **while** statement.

(D) Insert the appropriate "false label" and "true label" in quadruples corresponding to assignment instructions—in this case, the two labels are identical, and we refer to them as the "next label". An assignment quadruple q is immediately followed by:

(1) a; occurring outside a quadruple, or
(2) an **end**, or
(3) the first blank after the text pf P_3.

In case (1) the "next label" of q is the first entry in the first quadruple

[2] **begin**'s and **end**'s are paired off in the same way as a well-formed sequence of left- and right-hand parentheses.

following the ; in question. In case (2), we first find the **begin** matching the **end** in question, and the "next label" of q is the first entry in the test quadruple immediately preceding this **begin**. In case (3), the "next label" of q is $n + 1$, where n is the total number of instructions in P. Call the outcome of this stage P_4.

(E) Insert the appropriate "false label" in quadruples corresponding to test instructions. To find the "false label" of a test quadruple q we continue the left-to-right scan of P_4 until we bypass a well-formed (possibly empty) sequence of **begin-end** pairs and reach:

(1) a; occurring outside a quadruple, or
(2) an **end**, or
(3) the first blank after the text of P_4.

Cases (1), (2), and (3) are treated exactly in the same way as in stage D. Call the outcome of this stage P_5.

(F) Insert the appropriate "true label" in test quadruples. The "true label" of a test quadruple q is the first entry in the first quadruple following q. The outcome of this stage is P_6.

(G) Scan the text of P_6 deleting all characters occurring outside quadruples.

This completes the description of the algorithm for *quad*. Its implementation as a **while**-program is carried out by programming each stage in terms of string-processing functions, which are themselves function subprograms in the language of **while**-programs, as discussed in Section 3.3. Some of the programming details are left to the reader (Exercise 3.3.5). □

By way of illustration let us apply the algorithm of the preceding proof to the text of the **while**-program Q shown above. The outcome of stage A in the case of Q is:

```
begin while (; X1 ≠ X2; ;) do
            begin(; X1 := succ(X1); ;);
            (; X3 := pred(X3); ;) end;
      while (; X2 ≠ X3; ;) do (; X1 := 0; ;)
end
```

The outcome of stages B and C is:

```
while (1; X1 ≠ X2; ;) do
      begin (2; X1 := succ(X1); ;); (3; X3 := pred(X3); ;) end;
while (4; X2 ≠ X3; ;) do begin (5; X1 := 0; ;) end
```

The outcome of stages D, E, and F, is:

```
while (1; X1 ≠ X2; 4; 2) do
      begin (2; X1 := succ(X1); 3; 3); (3; X3 := pred(X3); 1; 1) end;
while (4; X2 ≠ X3; 6; 5) do begin (5; X1 := 0; 4; 4) end
```

Finally the outcome of stage G is again the list of quadruples mentioned before Proposition 4.

At this point we need to introduce the convention that if α is a string of symbols, then 'α' refers to the numerical code of α. The instruction

$$X := \text{`}\alpha\text{'}$$

is therefore a short-hand notation for

$$X := \text{"the numerical code of } \alpha\text{"}.$$

For example, $X :=$ '**begin while**' means $X := 2082$. This kind of assignment instruction is valid so long as α is a sequence of symbols that have been assigned codes in accordance with Table 1 of Section 3.1.

In the construction of the universal **while**-programs below, we also use three special-purpose functions, called *max*, *fetch*, and *next*. Suppose e is the index (numerical code) of a syntactically correct **while**-program P, and the value of $quad(e)$ is stored in variable LIST. We define *max* as follows:

$$max(\text{LIST}) = \text{"numerical code of largest label in LIST"}$$

i.e., if P has n instructions, then $max(\text{LIST}) = \text{`}n+1\text{'}$. In other words, $max(\text{LIST})$ will be the exit address for $quad(e)$.

If variable LBL = 'l', where l is some label $1 \leqslant l \leqslant n+1$, we then define

$$fetch(\text{LIST, LBL}) = \begin{array}{l}\text{"numerical code of the second}\\ \text{entry (the instruction) in a qua-}\\ \text{druple of LIST whose 1st entry is}\\ \text{LBL".}\end{array}$$

And if variable BLN holds '0' or '1' (corresponding to Boolean values *false* and *true*), we define:

$$next(\text{LIST, LBL, BLN}) = \begin{array}{l}\text{"numerical code of the third or}\\ \text{fourth entry (a label) in the qua-}\\ \text{druple of LIST whose first entry}\\ \text{is LBL, according to whether}\\ \text{BLN is 0 or 1".}\end{array}$$

We leave it to the reader to prove that *max*, *fetch*, and *next* are computable by **while**-programs (Exercise 6 of Section 3.3).

In the next two sections we justify our use of string-processing and prove Propositions 2 and 3. However, with these results (if not their proofs) at hand, we are now ready to construct our interpreter.

PROOF OF THE ENUMERATION THEOREM. We define a **while**-program, *interpreter*, that computes the universal functions $\Phi : \mathbf{N}^{j+1} \to \mathbf{N}$. The input variables of our interpreter are $E, X1, X2, \ldots, Xj$. Variable E is intended to hold code e of program $\mathbf{P}_e$. If r is the "extra-variable" factor computed in Proposition 3, we pose $k = j + r$. In addition to the input variables, the interpreter mentions variables $X(j+1), \ldots, Xk$, LIST, LBL, MAX, BLN, and INST. (We use variable names of a form different from Xn for convenience and clarity.) Our universal **while**-program also uses function subprograms *legal*, *short*, *quad*, *max*, *fetch*, and *next*, as well as some of the macro statements defined in Section 2.2.

The interpreter given below starts by checking if the value stored in variable E encodes a syntactically correct **while**-program—if it does not, the computation diverges (first line in the program). But if the value in E is the numerical code of a bona-fide **while**-program, the interpreter computes the numerical code of the *short*-version of the latter (second line in the program) which uses k variables, and then finds its sequence of quadruples (third line in the program). The rest of the program is a self-explanatory simulation of the *short*-version of the **while**-program whose index is stored in E. It simply keeps simulating the next instruction until it reaches the exit (whose address is stored in MAX). Variable INST holds the code of the instruction to be simulated next (it is the "instruction register"); while LBL holds the address of the instruction to be simulated next (it is the "instruction counter").

```
begin while legal(E) ≠ 1 do begin end;
      E := short(E);
      LIST := quad(E);
      LBL := '1';
      MAX := max(LIST);
      while LBL ≠ MAX do
            begin INST := fetch(LIST, LBL); BLN := '0';
                  if INST = 'X1 := 0' then X1 := 0;
                  ⋮
                  if INST = 'Xk := 0' then Xk := 0;
                  if INST = 'X1 := succ(X1)' then X1 := succ(X1);
                  if INST = 'X1 := pred(X1)' then X1 := pred(X1);
                  if INST = 'X1 := succ(X2)' then X1 := succ(X2);
                  if INST = 'X1 := pred(X2)' then X1 := pred(X2);
                  ⋮
                  if INST = 'Xk := succ(Xk)' then Xk := succ(Xk);
                  if INST = 'Xk := pred(Xk)' then Xk := pred(Xk);
                  if INST = 'X1 ≠ X2' ∧ X1 ≠ X2 then BLN := '1';
                  if INST = 'X1 ≠ X3' ∧ X1 ≠ X3 then BLN := '1';
                  ⋮
                  if INST = 'X(k − 1) ≠ Xk' ∧ X(k − 1) ≠ Xk then BLN
                      := '1';
                  LBL := next(LIST, LBL, BLN)
            end
end
```

Note that we set BLN to 0 as the default value on each cycle of the simulation, and that we only change it to 1 if we simulate a test, and the test returns *true*. We need not test $Xi \neq Xi$? which always returns *false*. □

To complete our study of the interpreter (universal program), we now turn, in the next three sections, to the study of string-processing, and the proofs of Propositions 2 and 3. Before doing so, however, we restate the Enumeration Theorem (for $j = 1$) in the following very useful form.

5 Proposition. *There is a macro statement, $X := \Phi(Y, Z)$, whose macro definition is a* **while**-*program with the following effect: Let Y and Z have values i and j, respectively, prior to execution. Then, if $\varphi_i(j)$ is undefined, execution of the macro will never terminate. Otherwise, it will terminate and X will then hold value $\varphi_i(j)$ while Y and Z will be unchanged unless they coincide with X.*

As with all other macros, we shall assume that wherever $X := \Phi(Y, Z)$ is called, its "hidden variables" are suitably relabelled to avoid clashes with variables in the surrounding text.

Exercises for Section 3.2

1. If $\Phi : \mathbf{N}^2 \to \mathbf{N}$ is the universal function for the set of all unary computable functions, how many indices does Φ have as a computable function? Justify your answer.

2. If $\mathbf{P}_e$ is a **while**-program that computes the universal function $\Phi : \mathbf{N}^2 \to \mathbf{N}$, show that the outcome of the computation of $\mathbf{P}_e$ on input vector (e, a) is identical to the outcome of its computation on input vector $(a, 0)$ for all $a \in \mathbf{N}$.

3. Let $f : \mathbf{N} \to \mathbf{N}$ be a computable bijection and consider the following alternative enumeration of **while**-programs: $P_0, P_1, \ldots, P_n, \ldots,$ where P_n is $\mathbf{P}_{f(n)}$ for all $n \in \mathbf{N}$. Prove that there is a computable universal function $\Psi : \mathbf{N}^2 \to \mathbf{N}$ for the resulting enumeration of unary computable functions: $\psi_0, \psi_1, \ldots, \psi_n, \ldots,$ where $\psi_n = \varphi_{f(n)}$. Show that Ψ is computed by a **while**-program $\mathbf{P}_{f(n)}$, for some n. Does the same hold if f is total and onto, but not necessarily one-to-one? What if f is total and one-to-one, but not onto?

4. Let $\Phi : \mathbf{N}^2 \to \mathbf{N}$ be the universal function for the set of unary computable functions. Show that the function $\theta(x) = \Phi(x, x)$ cannot be extended[3] to a total computable function. Deduce the existence of partial computable functions which do not have total computable extensions. (*Hint*: If θ' were a total computable extension of θ, then $\theta''(x) = \theta'(x) + 1$ would also be computable and total, contradicting the fact that for all $i \in \mathbf{N}$, $\Phi(i, x) \neq \theta''(x)$.)

5. The unsolvability of the Halting Problem says: There is no algorithm for deciding if an arbitrary computable function φ_i is defined on its own index i as argument. Prove the following unsolvability result: There is a single computable function φ_i for which there is no algorithm that can decide if $\varphi_i(x)$ is defined for an arbitrary argument $x \in \mathbf{N}$. (*Hint*: Derive this φ_i from the universal function for the family of computable unary functions.)

[3] We say ψ *extends* θ if $\theta \leqslant \psi$, that is, if whenever $\theta(x)$ is defined so too is $\psi(x)$, and $\psi(x) = \theta(x)$.

6. Let us redefine the concept of a "universal function" in a somewhat more general setting. Given a family $\mathfrak{F}$ of j-ary functions, which may or may not be computable, a *universal function* $F:\mathbf{N}^{j+1}\to\mathbf{N}$ for $\mathfrak{F}$ is characterized by two conditions:

   (a) For every fixed number e, the j-ary function $F(e,x_1,\ldots,x_j)$ belongs to $\mathfrak{F}$.
   (b) For every function $f(x_1,\ldots,x_j)$ in $\mathfrak{F}$, there exists a number e such that for all $x_1,\ldots,x_j$:
   $$F(e,x,\ldots,x_j)=f(x_1,\ldots,x_j).$$

   For simplicity, we assume here that $j=1$.

   (a) What is a universal function of the family of functions $\mathfrak{F}=\{1,x,x^2,x^3,x^4,\ldots\}$?
   (b) What is the universal function of the family $\mathfrak{F}=\{x,x^3,x^5,x^7,x^9,\ldots\}$?
   (c) Show that any finite family $\mathfrak{F}$ of functions has a universal function F.

7. We use the definition of "universal function" given in the preceding exercise.

   (a) Show that if $\mathfrak{F}$ is a family of total unary functions, which has a universal function $F:\mathbf{N}^2\to\mathbf{N}$, then there is a total unary function which does not belong to $\mathfrak{F}$ (*Hint*: Use a Cantor diagonalization.)
   (b) Show that the family of all unary functions on $\mathbf{N}$, which are both total and computable, do not have a *computable* universal function.
   (c) Define an infinite family of total computable functions from $\mathbf{N}$ to $\mathbf{N}$ which have a computable universal function. Display both the family and its universal function.

8. (a) Prove the existence of a computable function $\psi:\mathbf{N}\to\mathbf{N}$ satisfying the following condition:
   For every computable function $\theta:\mathbf{N}\to\mathbf{N}$, there is a total computable function $f:\mathbf{N}\to\mathbf{N}$ such that for all $x\in\mathbf{N}:\psi(f(x))=\theta(x)$.
   (b) Exhibit a computable function $\psi:\mathbf{N}\to\mathbf{N}$ which does not satisfy the condition given in (a).
   (c) Prove the existence of infinitely many computable functions $\psi:\mathbf{N}\to\mathbf{N}$ satisfying the condition given in (a).

3.3 String-Processing Functions

Below we describe three standard string-processing functions, where α and β are arbitrary character strings:

(1) *head* is applied to one argument α and returns the left-most symbol in α if α is not empty, otherwise it returns the empty string;
(2) *tail* is applied to one argument α and returns α with its left-most symbol chopped off if α is not empty, otherwise it returns the empty string;
(3) $\|$ is applied to two arguments α and β, and "concatenates" them in this order; i.e., it returns a string formed by appending β to the right of α.

If we now restrict all character strings to be over the basic character set of Table 1, then we can interpret any string as a natural number. We now define a number-theoretic function, also called *head*, which on argument $n \in \mathbf{N}$ returns the integer represented by the high-order six bits in the binary representation of n in case its binary representation has more than six bits. For smaller n, we arbitrarily set $head(n) = n$.

Likewise, number-theoretic $tail(n)$ returns the number obtained by chopping off the high-order six bits in the binary representation of n for any $n \in \mathbf{N}$ whose binary representation has more than six bits. Otherwise $tail(n) = 0$.

The value of $m \| n$ for any $m, n \in \mathbf{N}$ and $n \neq 0$, is the number obtained by writing the binary representation of m first, and immediately after it that of n. If $n = 0$ we agree that $m \| n = m$.[4] (Warning: We may get a different value for $m \| n$ if we write m and n in decimal; for example, $2 \| 4 = 20$ and not 24.)

In the case where n is the numerical code for a legal (non-empty) string α over our basic character sets, it is easy to see that the value of $head(n)$ is indeed the code for the first symbol in α, $tail(n)$ is the code of α with the first symbol removed, and $head(n) \| tail(n)$ is exactly n. If n is not the code of a legal string, *head*, *tail*, and $\|$, as defined above, are not always well-behaved functions. For example, if $n = 10000001$ then

$$head(n) \| tail(n) = 1000001 \neq n.$$

For the purpose of illustration, let α be the following string:

$$\mathbf{begin}\ X1 := succ(X1)\ \mathbf{end}$$

The numerical code of α is denoted by 'α' (which here is some number between 2^{59} and 2^{60}, since 'α' in binary has 60 bits). As explained above, we can use *head*, *tail*, and $\|$, as string-processing functions or as number-theoretic functions. We then have

$$\text{`}head(\alpha)\text{'} = head(\text{`}\alpha\text{'}) = \text{`}\mathbf{begin}\text{'},$$

$$\text{`}tail(\alpha)\text{'} = tail(\text{`}\alpha\text{'}) = \text{`}X1 := succ(X1)\ \mathbf{end}\text{'},$$

$$\text{`}head(\alpha)\text{'} \| \text{`}tail(\alpha)\text{'} = head(\text{`}\alpha\text{'}) \| tail(\text{`}\alpha\text{'}) = \text{`}\alpha\text{'}.$$

(Our standing assumption is that **begin**, **end**, *succ*, etc., are "undissectable" blocks; i.e., each is viewed as a single symbol.) Only the number-theoretic versions of *head*, *tail*, and $\|$, can formally appear in **while**-programs, which may only use data type "non-negative integer".

Exercise 1 at the end of this section asks for a proof that, as number-theoretic functions, *head*, *tail*, and $\|$ are computable by **while**-programs. This means that any string-processing function programmable in terms of *head*, *tail*, and $\|$ can also be viewed as a computable number-theoretic function.

[4] Our definition of (the number-theoretic versions of) *head*, *tail*, and $\|$ implicitly makes zero the numerical code of the empty string.

SYNTACTIC ANALYSIS

We now prove Proposition 2 of Section 3.2.

1 Proposition (Proposition 3.2.2). *There is an effectively computable function* $legal: \mathbf{N} \to \{0, 1\}$ *such that*

$$legal(e) = \begin{cases} 1, & \textit{if e encodes a syntactically} \\ & \textit{correct } \textbf{while}\textit{-program;} \\ 0, & \textit{otherwise.} \end{cases}$$

PROOF (outlined)[5]. We write a **while**-program P that computes function *legal*. Following our plan for encoding **while**-programs in Section 3.1, P goes through three stages—in this order:

(A) Check whether the length of the binary representation of e is a multiple of 6.
(B) Check whether every block of six bits in the binary representation of e is one of the binary codes listed in Table 1.
(C) Check whether e is the numerical code of a syntactically correct **while**-program.

Whenever any of the stages above fails, program P aborts the computation and outputs 0; if all stages succeed, P outputs 1. We can further specify the operation of stage C according to the following algorithm, where variable E initially holds value e, and α and β are arbitrary nonempty strings over the basic character set of **while**-programs:

1. If $E =$ '**begin** α **end**' then set E to 'α' and go to 2, or else output 0 and STOP.
2. If $E =$ "code of a single assignment statement" then output 1 and STOP, or else go to 3.
3. If $E =$ '**begin** α **end**' then set E to 'α' and go to 2, or else go to 4.
4. If $E =$ '**begin** α **end**; β' (with the sequence of **begin**'s and **end**'s in α well-formed) then set E to 'α; β' and go to 2, or else go to 5.
5. If $E =$ '**while** $Xi \neq Xj$ **do** α' (for some $i, j \in \mathbf{N}$) then set E to 'α' and go to 2, or else go to 6.
6. If $E =$ '$Xi := 0$; α' (for some $i \in \mathbf{N}$) then set E to 'α' and go to 2, or else go to 7.
7. If $E =$ '$Xi := succ(Xj)$; α' (for some $i, j \in \mathbf{N}$) then set E to 'α' and go to 2, or else go to 8.
8. If $E =$ '$Xi := pred(Xj)$; α' (for some $i, j \in \mathbf{N}$) then set E to 'α' and go to 2, or else go to 9.
9. Output 0 and STOP.

[5] For readers familiar with context-free languages, a more direct proof is available: The set of syntactically correct **while**-programs is context-free and, hence, there is an algorithm to check whether or not a string over the basic character set of **while**-programs belongs to the set.

The above algorithm is based on the inductive definition of **while**-programs in Section 2.1. This algorithm is not "quite correct", however, because it returns 0 whenever e encodes a syntactically correct **while**-program containing occurrences of the pair "**begin end**" (without any text in between). To handle such occurrences we can either add appropriate steps to the algorithm, or change its input e to e' which now encodes a string where every pair "**begin end**" is replaced by "**begin** $X1 := succ(X1)$; $X1 := pred(X1)$ **end**". We have not included the required changes in the algorithm for the sake of simplicity. □

Implementation of all programming steps in the preceding proof as well as in the proof of Proposition 4 or Section 3.2 can be carried out in terms of the string-processing functions *head*, *tail*, and || (Exercises 3, 4, and 5 at the end of this section). This is no coincidence. It turns out that any string-manipulation procedure is programmable in terms of *head*, *tail*, and ||, as discussed next.

An alternative framework for computability theory

We have developed our theory by working at functions on the domain **N** of natural numbers, which are programmable in terms of four basic operations (given "for free"), namely *succ*, *pred*, "writing down zero", and "testing whether two natural numbers are equal". This has been our characterization of the computable functions. The careful reader will have noticed that our results so far, except for the arithmetization of **while**-programs in Section 3.1, are nowhere bound to a specific system of notation for the natural numbers. But even the binary representation of natural numbers, which we used in the arithmetization of **while**-programs, was chosen somewhat arbitrarily. It could be replaced by another system of notation for **N** (Exercise 1 in Section 3.1), or altogether eliminated (Exercise 2 in Section 3.1). We can thus develop all of computability theory outside the framework of any privileged system of notation for the natural numbers.[6]

Alternatively we can start with a fixed finite alphabet Σ, and then study computability on the domain Σ^* of all finite strings over Σ. We then say that the function on Σ^* is computable if it is programmable in terms of a few basic operations on Σ^* which are given "for free"—most commonly these operations are *head*, *tail*, ||, "assigning the value α for any $\alpha \in \Sigma \cup \{\lambda\}$", and "testing whether a string is equal to α for any $\alpha \in \Sigma \cup \{\lambda\}$". ($\lambda$ denotes the empty string.) Readers familiar with the programming language

[6] In Section 6.2 we also show that all of computability theory—save for a few technical results—can be made model independent in addition to being notation independent. That is, almost all of computability theory can also be developed outside the framework of any privileged "model of computation" or programming language, such as **while**-programs.

LISP will recognize that these operations are respectively CAR, CDR, CONS, "assigning an atom or NIL", and "testing whether an s-expression is an atom or NIL". The language of **while**-programs appropriately modified for the study of computability on Σ^* is given next.

2 Definition. Σ is a fixed finite alphabet, and the operations *head*, *tail*, and $\|$, on Σ^* are given "for free". The corresponding language of **while**-programs is defined inductively by:

(1) An *assignment statement* is in one of the following forms:

$$X := \alpha, \text{ for } \alpha \in \Sigma \cup \{\lambda\},$$

$$X := head(Y),$$

$$X := tail(Y),$$

$$X := Y \| Z,$$

where X, Y and Z are arbitrary variable names.

(2) A **while** *statement* is of the form:

$$\textbf{while } X \neq \alpha \textbf{ do } \mathbb{S}$$

where $\alpha \in \Sigma \cup \{\lambda\}$ and $\mathbb{S}$ is an arbitrary statement.

(3) A *compound statement* is of the form:

$$\textbf{begin } \mathbb{S}_1; \ldots ; \mathbb{S}_m \textbf{ end}$$

where $m \geqslant 0$ and $\mathbb{S}_1, \ldots \mathbb{S}_m$ are arbitrary statements. As before, a **while**-*program* is simply a compound statement.

In this alternative framework, Church's Thesis asserts: Any partial function on Σ^* which is in one way or another algorithmically defined is also computable by a **while**-program (as defined above). Note that Σ is an arbitrary finite alphabet, e.g., Σ may be $\{0, 1\}$ or the basic character set of **while**-programs. In the latter case, a formal arithmetization of the **while**-programs is unnecessary, because **while**-programs and the data they operate on belong to the same domain Σ^*—and we can take the index of a **while**-program $P \in \Sigma^*$ to be P itself.

Once the syntax is arithmetized, i.e., once every string in Σ^* is assigned a number in $\mathbf{N}$, Exercise 1 at the end of this section shows that *head*, *tail*, and $\|$ (and therefore all effectively defined string-processing functions) are programmable as number-theoretic functions. Conversely, using an appropriate representation of the natural numbers by strings in Σ^*, Exercise 7 at the end of this section shows that all computable number-theoretic functions can be viewed as programmable string-processing functions.

Exercises for Section 3.3

1. Write **while**-programs that compute *head*, *tail*, and $\|$, as the number-theoretic functions defined at the beginning of this section.

2. Use the enumeration of **N** given in Exercise 1 of Section 3.1 to redefine the number-theoretic versions of *head*, *tail*, and $\|$.

3. Using *head*, *tail*, and $\|$ as function subprograms, write **while**-programs for the following string-processing functions:

   (a) $pre(\text{`}\alpha\text{'}, \text{`}\beta\text{'}) = \begin{cases} 0, & \text{if } \alpha \text{ is not a prefix of } \beta; \\ 1, & \text{if } \alpha \text{ is a prefix of } \beta. \end{cases}$

   (b) $suf(\text{`}\alpha\text{'}, \text{`}\beta\text{'}) = \begin{cases} 0, & \text{if } \alpha \text{ is not a suffix of } \beta; \\ 1, & \text{if } \alpha \text{ is a suffix of } \beta. \end{cases}$

   (c) $length(\text{`}\alpha\text{'})$ = number of characters in string α.
   (d) $del(n, \text{`}\alpha\text{'}) = \text{`}\beta\text{'}$, where β is the string obtained by deleting the first $n \in \mathbf{N}$ characters from α, if $length(\text{`}\alpha\text{'}) > n$—otherwise $del(n, \text{`}\alpha\text{'}) = 0$.
   (e) $sub(\text{`}a\text{'}, \text{`}b\text{'}, \text{`}\alpha\text{'}) = \text{`}\beta\text{'}$, where β is the substring of α from the first occurrence of character a (exclusive) leftward to the first occurrence of character b (exclusive), if characters a and b occur in α in this order—otherwise $sub(\text{`}a\text{'}, \text{`}b\text{'}, \text{`}\alpha\text{'}) = 0$.
   (f) $rot(n, \text{`}\alpha\text{'})$ which "rotates" string α n times is best defined recursively:

   $$rot(0, \text{`}\alpha\text{'}) = \text{`}\alpha\text{'},$$

   $$rot(n+1, \text{`}\alpha\text{'}) = tail(rot(n, \text{`}\alpha\text{'})) \| head(rot(n, \text{`}\alpha\text{'})).$$

   In the above definitions, α and β are arbitrary strings over the basic character set of **while**-programs. Recall that our conventions make zero the numerical code of the empty string. Note also that the functions defined above are not necessarily total number-theoretic functions. (Why?)

4. Consider **while**-program P in the proof of Proposition 1.

   (1) Write the parts of P that implement stages A and B.
   (2) Simulate a computation of the algorithm for stage C of P on input 'Q', where Q is the **while**-program in Example 5 of Section 2.3.
   (3) Write a **while**-program that implements stage C of P. For simplicity, assume that **while**-programs do not contain occurrences of the pair "**begin end**", without any text in between.

   (*Hint*: It may be useful to use some of the string-processing functions defined in the preceding exercise.)

5. Write a **while**-program for each of the seven stages of the algorithm in the proof of Proposition 4 of Section 3.2. (*Hint*: You may use some of the string-processing functions defined in Exercise 3, as function subprograms.)

6. Write **while**-programs to compute functions *max*, *fetch*, and *next* as defined before the proof of the Enumeration Theorem in Section 3.2. (*Hint*: Use some of the function subprograms of Exercise 3.)

7. Consider the language of **while**-programs as given in Definition 2 of this section, with $\Sigma = \{0, 1\}$. Assuming the binary representation of natural numbers, write **while**-programs for *succ* and *pred* (viewed as string-processing functions on Σ^*). Conclude that every computable number-theoretic function can be programmed as a string-processing function on Σ^*.

3.4 Pairing Functions

In this section we prove that we can effectively encode a vector of numbers into a single number, and then effectively decode the latter into its original components.

In Section 1.3, we met Cantor's diagonal argument that the reals, **R**, could *not* be put in bijective correspondence with the natural numbers, **N**. In another use of diagonals, Cantor contradicted the naive intuition that there are more rational numbers than natural numbers, by showing that the set $\{x/y \mid x, y \in \mathbf{N}, y > 0\}$ of all positive fractions can be *enumerated* (put in one-to-one correspondence with **N**) as shown in Figure 1. Let us call D_n the diagonal of all pairs x/y such that $x + y = n$. The enumeration then proceeds by counting through each diagonal $D_1, D_2, \ldots, D_n, D_{n+1}, \ldots$ in turn, in the downward order of increasing x for each diagonal, as shown in Figure 1(a).

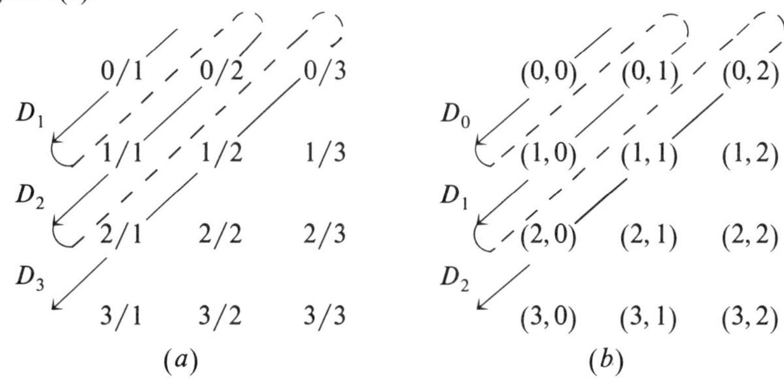

Figure 1 (a) Cantor's enumeration of the positive rationals. (b) Its adaptation to the enumeration of $\mathbf{N}^2$.

A pairing function is a computable bijection from $\mathbf{N}^2$ to **N** which provides an effective enumeration of pairs of integers. Although there are many different ways of defining specific pairing functions, we choose a particular map, here called τ, and keep it fixed throughout the book. We obtain τ by the simple adaptation of Cantor's enumeration shown in Figure 1(b).

1 Lemma. *The map* τ *given by*

$$\tau(i, j) = \tfrac{1}{2}(i + j)(i + j + 1) + i,$$

for all $i, j \in \mathbf{N}$, *is a computable one-to-one correspondence from* $\mathbf{N}^2$ *onto* **N**.

PROOF. τ is clearly computable; in particular, it is polynomial, and hence computable by a **while**-program by Proposition 1 of Section 2.2. Furthermore, the value assigned by τ to an ordered pair of natural numbers is its order of occurrence in Figure 1(b). To prove this, note first that there is exactly one pair whose components sum up to 0, two pairs whose components sum up to 1, . . . , and, in general, exactly k pairs whose components sum up to $(k-1)$. Hence, there are exactly $1+2+\cdots+k=\frac{1}{2}k(k+1)$ pairs whose components sum up to $(k-1)$ or less. Recall that $(0,0)$ corresponds to 0 in Figure 1(b). Hence, if τ establishes the correspondence described above, the pairs $(0,k)$, $(1,k-1)$, $(2,k-2)$, . . . , $(k,0)$ must be respectively mapped to $\frac{1}{2}k(k+1)$, $\frac{1}{2}k(k+1)+1$, $\frac{1}{2}k(k+1)+2, \ldots,$ $\frac{1}{2}k(k+1)+k$. This means that, in general, the pair (i,j), such that $i+j=k$, must be mapped to $\frac{1}{2}(i+j)(i+j+1)+i$—which is indeed what $\tau(i,j)$ was defined to be. □

The effective encoding of two natural numbers i and j into $\tau(i,j)$ will be of use to us if we can also effectively decode $\tau(i,j)$ into its two components.

2 Lemma. *Define two projection functions, $\pi_1 : \mathbf{N} \to \mathbf{N}$ and $\pi_2 : \mathbf{N} \to \mathbf{N}$, as follows*:

$$\pi_1(\tau(i,j)) = i,$$

$$\pi_2(\tau(i,j)) = j,$$

for all $i, j \in \mathbf{N}$. Then π_1 and π_2 are total computable functions.

PROOF. First, it is easily checked that the pairing function τ is strictly monotone in each of its arguments; that is,

$$\tau(i,j) < \tau(i,j+1),$$

$$\tau(i,j) < \tau(i+1,j).$$

In particular, $i \leqslant \tau(i,j)$ and $j \leqslant \tau(i,j)$. This fact allows us to write the following macro definition for $X2 := \pi_1(X1)$; $X3 := \pi_2(X1)$ in order to compute the functions π_1 and π_2 simultaneously:

```
begin X2 := 0; X3 := 0;
      while τ(X2,X3) ≠ X1 do
            begin X3 := succ(X3);
                  if τ(X2,X3) > X1 then
                  begin X2 := succ(X2); X3 := 0 end
            end
end
```

Starting with the pair $(0,0)$, this program stores in $(X2,X3)$ consecutive pairs from the first row in Figure 1(b), until it finds a pair $(0,j)$ for which $\tau(0,j) > X1$. At this point, it starts enumerating consecutive pairs from the second row until it finds $(1, j)$ for which $\tau(1, j) > X1$. And then the program switches to the third row, and so on. Thus the program consecutively "sweeps" every row from left to right, stopping at the ith row whenever it finds a pair (i, j) for which $\tau(i, j) > X1$. Since τ is a strictly monotone function, this means that in the "sweep" of row i, every pair (i, j) such that $\tau(i, j) \leqslant X1$ is enumerated. Hence all pairs (i, j) in all rows such that $\tau(i, j) \leqslant X1$ are eventually enumerated, i.e., consecutively stored in $(X2, X3)$. Since τ is onto and strictly monotone, the procedure eventually encounters a pair of values for $(X2, X3)$ such that $\tau(X2, X3) = X1$.

We have thus shown that there is a macro definition for $X2 := \pi_1(X1)$; $X3 := \pi_2(X1)$ which is the **while**-program above, with input variable $X1$, and which converges for all possible input values. □

For an alternative proof of Lemma 2, see Exercise 4 at the end of this section, which gives closed-form expressions for the projection functions π_1 and π_2.

We now define computable bijections $\tau_k : \mathbf{N}^k \to \mathbf{N}$, one for each $k \geqslant 1$, as follows:

$$\tau_1(i_1) = i_1$$

and

$$\tau_k(i_1, \ldots, i_k) = \tau(\tau_{k-1}(i_1, \ldots, i_{k-1}), i_k),$$

for all $i_1, \ldots, i_k \in \mathbf{N}$. ($\tau_2$ is none other than τ.) We likewise define corresponding projection functions $\pi_{k1}, \ldots, \pi_{kk}$, for all $k \geqslant 1$, as follows:

$$\pi_{k1}(\tau_k(i_1, \ldots, i_k)) = i_1,$$

$$\vdots$$

$$\pi_{kk}(\tau_k(i_1, \ldots, i_k)) = i_k.$$

(π_{21} and π_{22} are none other than π_1 and π_2.)

By Lemma 1, the pairing function τ is computed by a **while**-program which uses m variables for some $m \in \mathbf{N}$. By the definition of τ_k, it is easy to see that τ_k can be computed by a **while**-program that uses no more than $m + k$ variables, for all $k \in \mathbf{N}$ (Exercise 1).

The fact that a fixed number of variables suffices to compute the projection function π_{kl}, for any k and l, is not trivial and in fact is somewhat surprising.

3 Lemma. *There is a fixed number n such that the projection function π_{kl}, for any $k \geqslant 1$ and $1 \leqslant l \leqslant k$, is computed by a* **while**-*program using no more than n variables.*

Proof. First note that there are **while**-programs for the functions π_1 and π_2 which use no more than $(3 + m)$ variables, where m is the number of variables in the **while**-program computing τ, as shown in the proof of the preceding lemma. Second, an arbitrary projection function π_{kl} is definable in terms of π_1 and π_2 as:

$$\pi_{kl}(i) = \begin{cases} \pi_1^{k-l}(i), & \text{if } l = 1; \\ \pi_2\left(\pi_1^{k-l}(i)\right), & \text{if } 2 \leqslant l \leqslant k; \end{cases}$$

where we use the notation π_1^{k-l} to indicate that the function π_1 is repeatedly applied $k - l$ times.

If P is the **while**-program constructed in the proof of Lemma 2 which uses $(3 + m)$ variables, the following **while**-program computes π_{kl}, with variable X_1 used for both input and output:

```
begin  P; X1 := X2;  ⎫
       P; X1 := X2;  ⎪
         ⋮           ⎬ (k − 1)
       P; X1 := X2   ⎭
end
```

The following **while**-program computes π_{kl} for $2 \leqslant l \leqslant k$:

```
begin  P; X1 := X2;  ⎫
       P; X1 := X2;  ⎪
         ⋮           ⎬ (k − l)
       P; X1 := X2;  ⎭
       P; X1 := X3
end
```

In both cases the required number of variables is $n = 3 + m$, which is independent of k and l. □

The next lemma says that every j-ary computable function can be computed by a $(j + r)$-variable **while**-program, where the "extra-variable" factor r is fixed over all j-ary computable functions. This result is not only important in its own right, but its proof will yield the proof of Proposition 3 of Section 3.2, which allowed us to bound the number of program variables needed in the definition of our $(j + 1)$-ary interpreter (universal function) of the j-ary computable functions.

4 Lemma. *There is an $r \in \mathbf{N}$ such that if $\varphi_P^{(j)} : \mathbf{N}^j \to \mathbf{N}$ is a j-ary function computed by* **while**-*program P, then we can construct a $(j + r)$-variable* **while**-*program P' such that $\varphi_P^{(j)} = \varphi_{P'}^{(j)}$.*

PROOF. Suppose the given program P uses a total of k variables: $X1$, $X2, \ldots, Xk$. Program P' will use the pairing function τ to encode the full set of k variables in P, and store value $\tau_k(X1, \ldots, Xk)$ in a special variable U—without using τ_k explicitly, however. Program P' will directly simulate P using U and the decoding functions π_{kl}, for $l \leqslant k$. By Lemma 3 the computation of π_{kl} requires a fixed number n of variables *independent of the values of k and l*. P' will therefore have j variables for input, m variables for τ, n variables for all π_{kl}, and variables U, V, and W (the latter two being auxiliary variables used to save temporary values)—thus proving the result. The value of r is $m + n + 3$, and thus independent of k. Below we describe the construction in more detail.

If $j > k$ we can let P' be P. In this case, since $P' = P$ is a k-variable program, it is also a j-variable as well as a $(j + r)$-variable program. The nontrivial case thus occurs when $j \leqslant k$. The final form of P' in this case will look as follows:

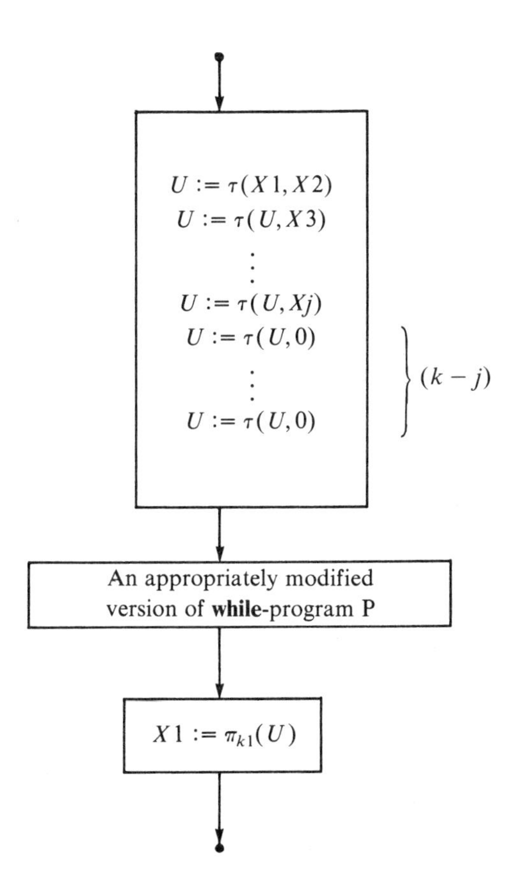

The first block in this diagram is a macro definition for $U := \tau_k(X1, \ldots, Xj, 0, \ldots, 0)$, which uses the full set of j input variables, the m variables for τ, and U. The appropriately modified version of P is obtained by replacing every assignment statement

$$Xi := g(Xj)$$

where $1 \leqslant i \leqslant k$ and g is *succ* or *pred* or 0, by the following sequence

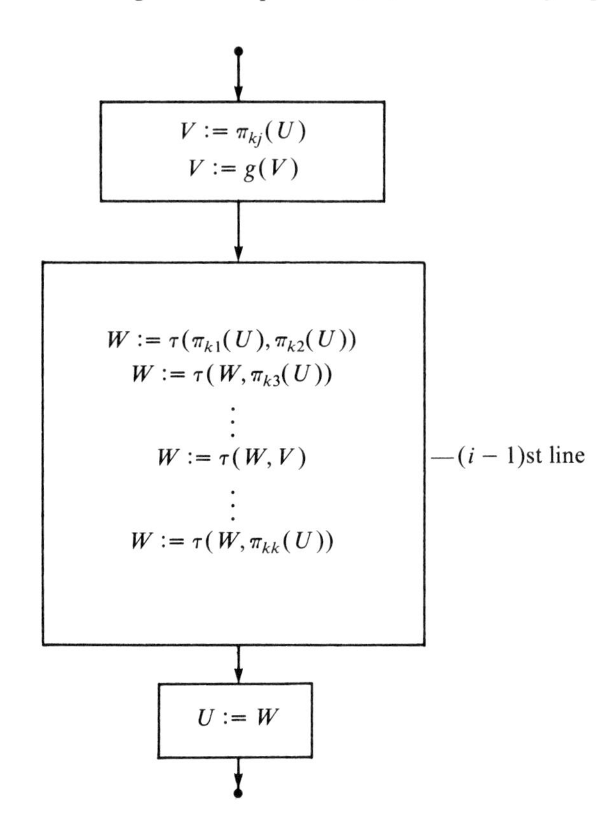

The second block in the diagram above has $(k - 1)$ lines, the $(i - 1)$st being $W := \tau(W, V)$. In case $i = 1$ the first line must be changed to $W := \tau(V, \pi_{k2}(U))$; and in case $i = 2$, to $W := \tau(\pi_{k1}(U), V)$. The variables used in the diagram above are U, V, W, the m variables for τ, and the n variables for $\pi_{k1}, \pi_{k2}, \ldots, \tau_{kk}$. Note that a crucial aspect in the construction of P' is our ability to use the same n variables in each of the function subprograms $\pi_{k1}, \pi_{k2}, \ldots, \pi_{kk}$—instead of using a distinct set of n variables in each one of them, which would result in a total of $k \times n$ variables.

Similarly, every **while** statement

$$\textbf{while } Xi \neq Xj \textbf{ do } \mathbb{S}$$

where $1 \leqslant i, j \leqslant k$ and $\mathbb{S}$ is an arbitrary statement is replaced by the sequence

$V := \pi_{ki}(U)$;
$W := \pi_{kj}(U)$;
while $V \neq W$ **do**
  begin $\mathbb{S}$; $V := \pi_{ki}(U)$; $W := \pi_{kj}(U)$ **end**

It is not too difficult to check that P' computes the same j-ary function as P.[7] □

5 Proposition (Proposition 3.2.3). *For all $j \geqslant 1$ there is a natural number r and a total computable function short* : $\mathbf{N} \to \mathbf{N}$ (*which depends on j*) *such that*:

(a) $\varphi_e^{(j)} = \varphi_{short(e)}^{(j)}$.
(b) *Program* $\mathbf{P}_{short(e)}$ *uses* $j + r$ *variables*.

Proof. The algorithm used to construct P' from P in the preceding lemma defines a string-processing function. By the correspondence established in Section 3.3 between strings and natural numbers, function *short* is the number-theoretic version of the transformation that takes P to P'; that is, if e is the index of P then $short(e)$ is the index of P'. We have, therefore, an algorithm for *short*. This algorithm can be rewritten as a **while**-program, by appropriate use of the basic string-processing functions (*head*, *tail*, and $\|$), in a manner similar to Exercise 4 in Section 3.3. □

Exercises for Section 3.4

1. Consider the pairing functions $\tau : \mathbf{N}^2 \to \mathbf{N}$ and $\tau_k : \mathbf{N}^k \to \mathbf{N}$, for some fixed $k > 2$.
   (a) Write a **while**-program P that computes τ.
   (b) If P in part (a) uses m variables, write a **while**-program Q with no more than $m + k$ variables that computes τ_k.
   (c) Let $j \leqslant k$ and define the function $\theta : \mathbf{N}^j \to \mathbf{N}$ as

   $$\theta(x_1, \ldots, x_j) = \tau_k(x_1, \ldots, x_j, \underbrace{0, \ldots, 0}_{(k-j)})$$

   for all $x_1, \ldots, x_j$. Write a **while**-program R with no more than $m + j$ variables that computes θ.

2. Prove the following statement: Given a j-ary computable function $\varphi_i^{(j)} : \mathbf{N}^j \to \mathbf{N}$, there is a unary computable function $\varphi_{\hat{i}} : \mathbf{N} \to \mathbf{N}$ such that $\varphi_i^{(j)} = \varphi_{\hat{i}} \circ \tau_j$ (τ_j is the j-ary pairing function defined above). Show that given index i, index $\hat{i}$ can be effectively computed. (*Hint*: Consider **while**-program P which computes $\varphi_i^{(j)}$, and derive from it the **while**-program $\hat{P}$ which computes $\varphi_{\hat{i}}$. A review of the proof of Lemma 4 may be helpful too.)

3. Prove that the function $\hat{\tau}(i, j) = 2^i(2j + 1) \dot{-} 1$, for all $i, j \in \mathbf{N}$, is a pairing function. Prove also that the corresponding projection functions are:

   $$\hat{\pi}_1(k) = exp(0, k + 1),$$

   $$\hat{\pi}_2(k) = \left(\frac{k + 1}{2^{exp(0,k+1)}} \dot{-} 1 \right) \mathbf{div}\ 2.$$

[7] It is useful to note that P and P' will not, in general, compute the same l-ary functions, for $l \neq j$.

The function **div** was defined in Proposition 1 of Section 2.2, and *exp* in Exercise 8 of Section 2.3.

4. Define the equality function $eq : \mathbf{N}^2 \to \mathbf{N}$ as follows:

$$eq(x, y) = 1 \dot{-} [(x \dot{-} y) + (y \dot{-} x)] = \begin{cases} 1, \text{ if } & x = y; \\ 0, \text{ if } & x \neq y. \end{cases}$$

(a) Verify that for all $i, k \in \mathbf{N}$,

$$\sum_{x=0}^{k} eq(\tau(i,x),k) = \begin{cases} 1, & \text{if there is a } j \text{ such that } \tau(i,j) = k; \\ 0, & \text{otherwise.} \end{cases}$$

(b) Using part (a) show that:

$$\pi_1(k) = (\mu i \leqslant k)\left[\sum_{x=0}^{k} eq(\tau(i,x),k) = 1\right]$$

$$\pi_2(k) = (\mu j \leqslant k)\left[\sum_{x=0}^{k} eq(\tau(x,j),k) = 1\right]$$

(Review the definition of "bounded minimization" in Exercise 9 of Section 2.3.)

(c) Based on the expressions in part (b), write **while**-programs that compute π_1 and π_2.

5. Review the definition of "universal function" in Exercise 6 of Section 3.2. Let $\mathcal{F}$ be an arbitrary family of functions from **N** to **N**, and $F : \mathbf{N}^2 \to \mathbf{N}$ a universal function for $\mathcal{F}$. We say that $\mathcal{F}$ *contains its own universal function*, if $\mathcal{F}$ contains a function $\theta : \mathbf{N} \to \mathbf{N}$ such that

$$\theta(\tau(x, y)) = F(x, y)$$

where τ is the pairing function defined above.

(a) Show that the family of all unary computable functions contains its own universal function.

(b) Let $\mathcal{F}$ be a family of total functions from **N** to **N** (not necessarily computable) which is closed under functional composition, and contains *succ* as well as the function $\hat{\tau}(x) = \tau(x, x)$. Show that $\mathcal{F}$ cannot contain its own universal function. (*Hint*: If $F(x, y)$ is the universal function of $\mathcal{F}$ and $\theta(\tau(x, y)) = F(x, y)$, prove that $succ(\theta(\hat{\tau}(x)))$ cannot belong to $\mathcal{F}$.)

6. We assume the terminology of Exercise 6 in Section 3.2. Let $\mathcal{F}$ be the family of all polynomial functions in one variable. Each such function is a total computable function from **N** to **N**.

(1) Show that $\mathcal{F}$ has a computable universal function $F : \mathbf{N}^2 \to \mathbf{N}$.

(2) Prove that the universal function F of $\mathcal{F}$ cannot be a polynomial function. (*Hint*: Use part (b) of the previous exercise for a short proof.)

CHAPTER 4

Techniques of Elementary Computability Theory

In this chapter we illustrate several techniques for judging a function's computational status. Sometimes we can prove that a function is computable by writing down a concrete **while**-program that calculates the function's values. Often this involves an appeal to the Enumeration Theorem, which permits us to use the interpreter macro

$$X := \Phi(Y, Z).$$

This macro has the following interpretation: Let x, y, z, be the values of X, Y, and Z, respectively, prior to execution. Then if $\varphi_y(z)$ is defined, the Φ macro will terminate, leaving value $\varphi_y(z)$ in X, with each of variables Y and Z unchanged unless it equals X. If $\varphi_y(z)$ is undefined, execution of $\Phi(Y, Z)$ will never terminate.

In many cases, producing a concrete program for calculating a function is impractical, and so demonstrating that a function is computable must rely on a more abstract tool: Church's Thesis. An appeal to Church's Thesis requires that an algorithm for calculating some function be supplied in informal outline form as justification that the function is computable. In every case, a program for the outlined process could be supplied by a sufficiently diligent programmer.

Parametrization is still another tool for helping us judge a function's computational status. Suppose $\theta(x, y)$ is a two-variable computable function. If we set $x = c$, where c is a constant, then $\theta(c, y) = \theta'(y)$ is also computable. Now we often wish to reason about computations involving θ', but to do so we need an index for it. The s-m-n or *parameterization* theorem allows us to obtain such an index effectively from e, an index for θ, and c.

So far, we have encountered a single example of a noncomputable function, the *halt* function of Sections 1.2 and 3.1, which was constructed using a diagonalization argument. In the last section of this chapter we give further examples of noncomputable functions which have been constructed by diagonalization and related techniques.

We begin this chapter with several examples that introduce the techniques described above. The analyses supplied in these examples constitute the very basis of computability theory and should be studied with great care.

4.1 Algorithmic Specifications

We now provide five examples which explore some of the subtleties of proofs of computability and noncomputability.

1 Example. Define the function ψ_1 by the specification

$$\psi_1(x, y) = \begin{cases} y, & \text{if } \varphi_x(x)\downarrow; \\ \perp, & \text{otherwise.} \end{cases}$$

To see that ψ_1 is computable, we need simply observe that it is computed by the following **while**-program, written using our interpreter macro. (We follow the convention that x and y will provide the initial values of $X1$ and $X2$, respectively, and that $\psi_1(x, y)$ is defined if and only if the computation terminates, and then with the required value in $X1$.)

$$\begin{aligned} &\textbf{begin} \quad X1 := \Phi(X1, X1); \\ &\qquad\quad\;\; X1 := X2 \\ &\textbf{end} \end{aligned}$$

The program terminates if and only if the interpreter terminates, i.e., if and only if $\varphi_x(x)\downarrow$; but then $X1 := X2$ sets the final result to y.

2 Example. Define the function ψ_2 by the specification

$$\psi_2(x, y) = \begin{cases} y, & \text{if } \varphi_x(x)\downarrow; \\ 0, & \text{otherwise.} \end{cases}$$

$\psi_2(x, y)$ is not computable. If it were, it would solve the Halting Problem—just fix y to take the value 1. In addition, notice that the behavior of $\mathbf{P}_x$ on x appears to be inconsistent with the "otherwise" clause of the specification, since a value is returned, namely 0, based on the condition that computation $\mathbf{P}_x$ on x runs forever. However, the next example shows that an apparent inconsistency with the "otherwise" clause need not, in fact, be one.

3 Example. Let $\sigma : \mathbf{N} \to \mathbf{N}$ and $\mu : \mathbf{N} \to \mathbf{N}$ be two computable functions such that $\sigma \leqslant \mu$. This means that whenever $\sigma(y)$ is defined, so is $\mu(y)$, and in this case we must have $\sigma(y) = \mu(y)$. Thus σ is a subset of μ when considered as a set of ordered pairs of natural numbers. Now define ψ_3 by the specification

$$\psi_3(x, y) = \begin{cases} \mu(y), & \text{if } \varphi_x(x)\downarrow; \\ \sigma(y), & \text{otherwise.} \end{cases}$$

At first glance, the second line asks us to return a value $\sigma(y)$ based on the assumption that computation $\mathbf{P}_x$ on x runs forever. But on closer inspection, we realize that $\sigma \leqslant \mu$ tells us that if $\sigma(y)$ is defined it equals $\mu(y)$. Thus we *can* compute $\psi_3(x, y)$ as follows: Start the computations of $\sigma(y)$ and of $\mathbf{P}_x$ on x at the same time. If a result $\sigma(y)$ is returned, halt—for then $\sigma(y) = \mu(y)$ is the desired answer, no matter whether or not $\mathbf{P}_x$ halts on x. But if $\mathbf{P}_x$ halts on x before a result $\sigma(y)$ is returned, stop computing $\sigma(y)$ and start computing $\mu(y)$. If and when this computation halts, the $\mu(y)$ then returned is the desired answer. Thus, by Church's Thesis $\psi_3(x, y) = \varphi_e^{(2)}(x, y)$ for some e. (A formal proof would spell out how to "interleave" the steps of the parallel computations of the above outline into a single serial **while**-program.)

4 Example. Define the function ψ_4 by the specification

$$\psi_4(x) = \begin{cases} 1, & \text{if the decimal expansion of } \pi = 3.14159265\ldots \\ & \text{has a block of consecutive 5's of length} \\ & \text{at least } x; \\ 0, & \text{otherwise.} \end{cases}$$

The reader may have already studied ψ_4 in Exercise 4 of Section 1.3, which was designed to reinforce the discussion of "Fermat's function" in Section 1.1. $\psi_4(x)$ *is* computable, although the specification given above does not establish this fact. Indeed, the specification suggests that π be expanded, and on argument x a block of x 5's be searched for. If no such block exists, then a nonterminating search will never turn up the sought-after string of 5's. Nevertheless, $\psi_4(x)$ *is* computable. We consider two cases. If there is a finite bound on the size of strings of 5's, say k, then $\psi_4(x)$ may be specified in an obviously algorithmic way by

$$\psi_4(x) = \begin{cases} 1, & \text{if } x \leqslant k; \\ 0, & \text{otherwise.} \end{cases}$$

On the other hand, if there is no finite bound on the size of blocks of consecutive 5's, then $\psi_4(x) \equiv 1$, and again ψ_4 is obviously computable. As in our "Fermat's last theorem" example of Section 1.1 we don't know which of infinitely many (in this case) algorithms applies, but we do know that there must be *some* algorithm which calculates the function in question.

5 Example. Define the function ψ_5 by the specification

$$\psi_5(x, y, z) = \begin{cases} 1, & \text{if } \mathbf{P}_x \text{ on input } y \text{ halts within } z \text{ steps;} \\ 0, & \text{otherwise.} \end{cases}$$

$\psi_5(x, y, z)$ is computable, since by simply running $\mathbf{P}_x$ on y for z steps[1] and checking to see if $\mathbf{P}_x$ has halted, we can evaluate the function. Obviously, $\psi_5(x, y, z)$ is also a total function. Notice that $\varphi_x(y)\downarrow$ if and only if $\psi_5(x, y, z) = 1$ for some z.

The number of steps taken to compute a function is an important measure of its *complexity*. Since *step counting* is an important measure, we show explicitly how to convert a **while**-program $\mathbf{P}_x$ into a new program $\mathbf{P}_w$ which computes

$$\theta(y, z) = \begin{cases} 1, & \text{if } \mathbf{P}_x \text{ on input } y \text{ halts within } z \text{ steps;} \\ 0, & \text{otherwise;} \end{cases}$$

i.e., we shall have $\varphi_w^{(2)} = \theta$. The idea of the program is simple. We use a counter, C, which is initially set to 0. We expand each instruction to test if C has yet reached the given value z: If it has, the next step would take us past z, and so we halt with 0 as result; otherwise, we increment C by 1 and take that next step. If we ever reach the exit without using up the allotted z steps, we set the result to 1. (Thus, the modified program checks steps, but does not return $\varphi_x(y)$.)

To simplify the specification of $\mathbf{P}_w$, we will use an instruction "**goto** *exit*" which will terminate computation. Then $\mathbf{P}_w$, which uses variables $X1$ and $X2$ for input, can be written as

begin $C := 0;\ T := X2;\ X2 := 0;$
  $\hat{\mathbf{P}}_x;$
  exit: **if** $C \leqslant T$ **then** $X1 := 1$ **else** $X1 := 0$
end

The first line in this program initializes the counter C, places the test value z in a new variable T, and sets $X2$ to 0 as required for $\mathbf{P}_x$ to evaluate $\varphi_x(y)$. The second line is an appropriately modified version $\hat{\mathbf{P}}_x$ of $\mathbf{P}_x$, which does not execute more than z steps. The third line sets variable $X1$ to 1 or 0, according to whether or not the computation of $\hat{\mathbf{P}}_x$ terminates by reaching its "normal exit" (i.e., without using **goto** *exit*).

The transformation that takes $\mathbf{P}_x$ to $\hat{\mathbf{P}}_x$ is defined inductively as follows. If $\mathbb{S}$ is an assignment statement then $\hat{\mathbb{S}}$ is

begin $\mathbb{S}$;
  $C := succ(C)$;
  if $C > T$ **then goto** *exit*
end

[1] Recall that a *step* is an elementary instruction, i.e., a "box" or a "diamond" in the flow diagram of a **while**-program, as defined in Chapter 2.

If $\mathbb{S}$ is a compound statement of the form **begin** $\mathbb{S}_1; \ldots; \mathbb{S}_m$ **end**, $\hat{\mathbb{S}}$ is

$$\textbf{begin}\ \hat{\mathbb{S}}_1; \ldots; \hat{\mathbb{S}}_m\ \textbf{end}$$

If $\mathbb{S}$ is a **while** statement of the form **while** $Xi \neq Xj$ **do** $\mathbb{S}_1$, then we must not only rewrite $\mathbb{S}_1$ as $\hat{\mathbb{S}}_1$, but must also increment the counter every time we test "$Xi \neq Xj$?", whether the result is true or false. Hence $\hat{\mathbb{S}}$ is, in this case,

begin while $Xi \neq Xj$ **do**
        begin $C := succ(C)$;
            if $C > T$ **then goto** *exit*;
            $\hat{\mathbb{S}}_1$
        end
    $C := succ(C)$;
    if $C > T$ **then goto** *exit*
end

Having defined the translation $\mathbb{S} \mapsto \hat{\mathbb{S}}$, we now have $\hat{\mathbf{P}}_x$, and therefore $\mathbf{P}_w$ too. In the next section (Example 4), we show how to go effectively from φ_x to an index w for the corresponding step-counting function, without explicitly constructing the required program.

Exercises for Section 4.1

1. (a) Define $\theta_1(i, j) = \begin{cases} 1, & \text{if } \varphi_i(k)\downarrow \text{ for some } k < j; \\ \perp, & \text{otherwise.} \end{cases}$

   Is θ_1 computable? Is θ_1 total?

   (b) Define $\theta_2(i, j) = \begin{cases} 1, & \text{if } \varphi_i(k)\downarrow \text{ for some } k > j; \\ \perp, & \text{otherwise.} \end{cases}$

   Is θ_2 computable? Is θ_2 total?

2. Define the function ψ by

$$\psi(x) = \begin{cases} \varphi_x(x) + 1, & \text{if } \varphi_x(x)\downarrow; \\ \perp, & \text{otherwise.} \end{cases}$$

   (a) Show that ψ is computable.
   (b) If i is an index for ψ, i.e., if $\psi = \varphi_i$, give a step-by-step account of the computation of $\mathbf{P}_i$ on input i.
   (c) Show that ψ cannot be extended to a total computable function; that is, there is no total computable function f such that $\psi \leqslant f$.

3. (Rogers) Let α and β be unary computable functions such that $A = \text{DOM}(\alpha)$, $B = \text{DOM}(\beta)$, and $A \cap B = \varnothing$.

   (a) Does there necessarily exist a computable function γ such that $\gamma(A) = \{0\}$ and $\gamma(B) = \{1\}$?

(b) Does there necessarily exist a total computable function f such that $f(A) = \{0\}$ and $f(B) = \{1\}$? (*Hint*: The answer is no. Use the construction of the preceding exercise restricted to the case of functions from $\mathbf{N}$ to $\{0, 1\}$.)

4. Discuss the computational status of each of the unary functions f_1, f_2, and f_3 below—according to whether or not it is the case that for all $i, j, k \in \mathbf{N} - \{0\}$ and all $n > 2: i^n + j^n \neq k^n$ (Fermat's last theorem).

$$(1)\quad f_1(n) = \begin{cases} 1, & \text{if there are } i, j, k \in \mathbf{N} - \{0\} \text{ such that } i^n + j^n = k^n; \\ \perp, & \text{otherwise.} \end{cases}$$

$$(2)\quad f_2(n) = \begin{cases} \perp, & \text{if there are } i, j, k \in \mathbf{N} - \{0\} \text{ such that } i^n + j^n = k^n; \\ 0, & \text{otherwise.} \end{cases}$$

$$(3)\quad f_3(n) = \begin{cases} 1, & \text{if there are } i, j, k \in \mathbf{N} - \{0\} \text{ such that } i^n + j^n = k^n; \\ 0, & \text{otherwise.} \end{cases}$$

4.2 The *s-m-n* Theorem

Let us recall how $\varphi_i^{(m+n)}(y_1, \ldots, y_m, z_1, \ldots, z_n)$ is computed. Program $\mathbf{P}_i$ is looked up and discovered to be a k-variable **while**-program. Variables $X1, \ldots, Xk$ are loaded with $y_1, \ldots, y_m, z_1, \ldots, z_n$ (extra variables are filled with zeroes, left over inputs are ignored), and then $\mathbf{P}_i$ is executed. If and when this computation terminates, then the value of $\varphi_i^{(m+n)}(y_1, \ldots, y_m, z_1, \ldots, z_n)$ is the value of the variable $X1$ at that time.

Now suppose we consider the following program, applied to input vector $(z_1, \ldots, z_n)$:

```
begin Xm + 1 := X1;
        ⋮
      Xm + n := Xn;
      X1 := y1;
        ⋮
      Xm := ym
end
```

This is meant to transform the state of variables $(X1, \ldots, Xm, Xm + 1, \ldots, Xm + n)$ from $(z_1, \ldots, z_n, 0, \ldots, 0)$ to $(y_1, \ldots, y_m, z_1, \ldots, z_n)$. Unfortunately, this program fragment has a bug in it which is revealed as soon as $m + 1 \geqslant n$, because then $Xm + 1 := X1$ overwrites the *old* $Xm + 1$ before it can be read into its new position by $X2m + 1 := Xm + 1$, etc. We thus have to replace the simple (but wrong!) program above by the more complicated (but correct!) program Q:

$$
\begin{array}{ll}
\textbf{begin} & Xm+n+1 := X1; \\
& \vdots \\
& Xm+n+n := Xn; \\
& X1 := y_1; \\
& \vdots \\
& Xm := y_m; \\
& Xm+1 := Xm+n+1; \\
& \vdots \\
& Xm+n := Xm+n+n; \\
& Xm+n+1 := 0; \\
& \vdots \\
& Xm+n+n := 0 \\
\textbf{end} &
\end{array}
$$

Now suppose that we consider the following program

$$\mathbf{P}_{i'} = \textbf{begin } Q; \mathbf{P}_i \textbf{ end}$$

Clearly, once we know $y_1, \ldots, y_m$ and i, we can effectively form this program and calculate its index, i'. Thus we can write

$$i' = s_n^m(i, y_1, \ldots, y_m),$$

where the function $s_n^m : \mathbf{N}^{n+1} \to \mathbf{N}$ is total and computable by Church's Thesis. Notice that neither program $\mathbf{P}_i$ nor program $\mathbf{P}_{s_n^m(i,y_1,\ldots,y_m)}$ is, in general, total. The s_n^m map merely takes a syntactic object (program i) and modifies its syntax in a systematic way by inserting a sequence of assignment statements at the beginning of program i.

Moreover it should be clear that

$$\varphi^{(n)}_{s_n^m(i,y_1,\ldots,y_m)}(z_1, \ldots, z_n) = \varphi_i^{(m+n)}(y_1, \ldots, y_m, z_1, \ldots, z_n).$$

This is so because the assignment block in program $\mathbf{P}_{s_n^m(i,y_1,\ldots,y_m)}$ loads variables $X1, \ldots, Xm+n$ with inputs in exact agreement with our semantic rules for computations by $\mathbf{P}_i$. (Make sure you understand this equation thoroughly before going on.) This result, the *s-m-n* Theorem, plays a fundamental role in computability theory. We state it formally below, using the notation s_n^m to indicate that a vector of m arguments is "pulled down" into the parameter list of the function s_n^m in index position, while n arguments remain as function arguments.

1 Theorem. (The *s-m-n* Theorem). *There exists a total computable function* $s_n^m : \mathbf{N}^{n+1} \to \mathbf{N}$ *such that*

$$\varphi^{(n)}_{s_n^m(i,y_1,\ldots,y_m)}(z_1, \ldots, z_n) = \varphi_i^{(m+n)}(y_1, \ldots, y_m, z_1, \ldots, z_n).$$

The s-m-n Theorem is also called the Parametrization Theorem, because it shows that an index for a computable function can be effectively calculated from the list of parameters (here, $y_1, y_2, \ldots, y_m$) on which it depends.

We might term Theorem 1 a *functional form* of parametrization: the syntax of $\mathbf{P}_i$ is modified so that its semantics as an $(m + n)$-ary function coincides with the semantics of $\mathbf{P}_{s_n^m(i, y_1, \ldots, y_m)}$ thought of as an n-ary function. But other parametrizations are also possible. For example, consider program $\mathbf{P}_e$ below:

$$\mathbf{P}_e = \begin{cases} \textbf{begin} \\ \quad \vdots \\ \qquad \textbf{while } Xk \neq Xj \textbf{ do} \\ \qquad Xi := succ(\,\textcircled{Xj}\,) \\ \quad \vdots \\ \textbf{end} \end{cases}$$

If we replace the circled occurrence of Xj by y, we obtain the program equality

$$\mathbf{P}_{f(e,y)} = \begin{cases} \textbf{begin} \\ \quad \vdots \\ \qquad \textbf{while } Xk \neq Xj \textbf{ do} \\ \qquad Xi := succ(y) \\ \quad \vdots \\ \textbf{end} \end{cases}$$

Here again the transforming function f is total computable, since as a function of y it modifies syntax in a systematic way. But this equality has no useful corresponding form of functional equivalence. Nevertheless parameter-fixing of this kind is very useful. We summarize the principle with the following theorem.

2 Theorem. (The s-m-n Theorem, Program Form).[2] *Let* $\mathbf{P}_e$ *be a program, and suppose* $y_1, \ldots, y_m \in \mathbf{N}$. *If we replace any* m *occurrences of any variables in the text of* $\mathbf{P}_e$ *(which do not occur on the left-hand sides of assignment statements) with* $y_1, \ldots, y_m$, *respectively, then the resulting program is* $\mathbf{P}_{f(e, y_1, \ldots, y_m)}$, *where* f *is a total computable function.*

When we consider a total computable function which is best understood as a transformation of program syntax, we will term such a function a

[2] The "s-m-n" of Theorem 1 refers to the letters of the name of the transforming function. Since there is no similar name for the function appearing in the program form of the theorem, our title for Theorem 2 is a slight abuse of terminology.

program-rewriting function. The functions s_n^m and f in Theorem 1 and Theorem 2 above are thus instances of program-rewriting functions.

If we fix a variable in a macro call, replacing, say, $X1 := \theta(X1, \ldots, Xj, \ldots, Xk)$ by $X1 := \theta(X1, \ldots, y, \ldots, Xk)$, where θ is a macro, we intend this to mean that all test and right-hand-side assignment occurrences of Xj in the macro expansion of θ are replaced by y before the macro text is inserted in the main program. (By the conventions of Section 2.2, we assume that $X1, \ldots, Xk$ may not appear on the left-hand side of any assignment instruction in the macro expansion of θ.) In what follows we shall refer to applications of either Theorem 1 or Theorem 2 as the s-m-n Theorem, even though traditionally the name "s-m-n Theorem" has been reserved for Theorem 1 alone.

3 Example. Given computable functions of one variable, say φ_a and φ_b, it is clear that their composite, $\varphi_a \circ \varphi_b$ is also computable. By Church's Thesis there is a $c \in \mathbf{N}$ for which

$$\varphi_c = \varphi_a \circ \varphi_b .$$

Now suppose we compose two different functions, say $\varphi_{a'}$ and $\varphi_{b'}$. In this case we get a $c' \in \mathbf{N}$ such that

$$\varphi_{c'} = \varphi_{a'} \circ \varphi_{b'} .$$

There is certainly a functional relationship here between the indices of the composed functions and the index of the resulting function, which we can express as

$$\varphi_{\mu(m,n)} = \varphi_m \circ \varphi_n$$

with $\mu(a,b) = c$ and $\mu(a',b') = c'$.

Is μ computable? Is μ total? From a programming point of view, the answer to both questions should be yes—the function $\mu(m,n)$ has as a value the index of the program obtained by concatenating the following three blocks of code:

(1) the **while**-program decoding of n, i.e., $\mathbf{P}_n$;
(2) a block of code setting all variables except $X1$ to 0;
(3) the **while**-program decoding of m, i.e., $\mathbf{P}_m$.

Moreover, the function $\mu(m,n)$ is also total (thus earning μ a Roman name, say h), since programs $\mathbf{P}_m$ and $\mathbf{P}_n$ are *not* executed in the process of calculating $\mu(m,n)$, but instead are merely decoded and reencoded as syntactic objects.

The s-m-n Theorem establishes this result in a much more general way. Define a function $\theta : \mathbf{N}^3 \to \mathbf{N}$ which for all m, n, and x, is:

$$\theta(m,n,x) = \varphi_m \circ \varphi_n(x).$$

By Church's Thesis θ is computable—apply $\mathbf{P}_n$ to x, and if and when this computation terminates, load the result into variable $X1$ of $\mathbf{P}_m$ and execute that program. Hence

$$\varphi_e(m,n,x) = \theta(m,n,x) = \varphi_m \circ \varphi_n(x),$$

where e is some fixed value. By the *s-m-n* Theorem, we have

$$\varphi_{s_1^2(e,m,n)}(x) = \theta(m,n,x) = \varphi_m \circ \varphi_n(x).$$

Setting $h(m,n) = s_1^2(e,m,n)$, we obtain our desired result,

$$\varphi_{h(m,n)} = \varphi_m \circ \varphi_n .$$

The function h is total and computable since s_1^2 is total and computable.

4 Example. Consider again function ψ_5 of Section 4.1:

$$\psi_5(x,y,z) = \begin{cases} 1, & \text{if } \mathbf{P}_x \text{ on input } y \text{ halts within } z \text{ steps;} \\ 0, & \text{otherwise.} \end{cases}$$

By Church's Thesis, ψ_5 is computable, and therefore $\psi_5 = \varphi_n^{(3)}$ for some index n. For a fixed program x, we want to show the existence of an index for the step-counting function θ—but this time not by explicitly constructing a program for it. Recall that

$$\theta(y,z) = \begin{cases} 1, & \text{if } \mathbf{P}_x \text{ on input } y \text{ halts within } z \text{ steps;} \\ 0, & \text{otherwise.} \end{cases}$$

Hence, keeping x fixed as a parameter:

$$\theta(y,z) = \psi_5(x,y,z) = \varphi_n^{(3)}(x,y,z) = \varphi_{s_2^1(n,x)}^{(2)}(y,z),$$

where the last equality follows from the *s-m-n* Theorem. This shows that program $s_2^1(n,x)$, $\mathbf{P}_{s_2^1(n,x)}$, computes the step-counting function θ of $\mathbf{P}_x$. As x ranges over $\mathbf{N}$, we can also effectively enumerate indices for all step-counting functions, namely: $s_2^1(n,0)$, $s_2^1(n,1)$, $s_2^1(n,2)$, ...

5 Example. Just as we showed in the preceding example that we can effectively enumerate indices for all step-counting functions, we can use the *s-m-n* Theorem again in effectively indexing certain families of computable functions. We know, for instance, that the function $f(y,x) = x ** y$ is computable by a **while**-program, say $\mathbf{P}_n$ (Exercise 1 of Section 2.2), so that $f = \varphi_n^{(2)}$. Hence, in order to find an effective indexing for the sequence of functions $\{1, x, x^2, x^3, x^4, \ldots\}$, we can write:

$$f(y,x) = \varphi_n^{(2)}(y,x) = \varphi_{s_1^1(n,y)}(x),$$

and the desired effective enumeration of indices is $\{s_1^1(n,0), s_1^1(n,1), s_1^1(n,2), s_1^1(n,3), \ldots\}$. Of course, an alternative but more tedious approach is to explicitly construct a sequence of programs, with one program for each function in $\{1, x, x^2, x^3, \ldots\}$.

6 Example. Consider the program

$$\mathbf{P}_e = \begin{cases} \textbf{begin} \\ \qquad X2 := succ(X1); \\ \qquad \textbf{while } X1 \neq X2 \textbf{ do} \\ \qquad\qquad X2 := pred(X2) \\ \textbf{end} \end{cases}$$

which computes $\varphi_e^{(2)}(u,v) = u$, a total function.

If we form

$$\mathbf{P}_{f(e,i)} = \begin{cases} \textbf{begin} \\ \qquad X2 := succ(i); \\ \qquad \textbf{while } X1 \neq X2 \textbf{ do} \\ \qquad\qquad X2 := pred(i) \\ \textbf{end} \end{cases}$$

which consists in replacing every variable on the right-hand side of an assignment by i, then $f(e,i)$ is total computable by the program form of the s-m-n Theorem. For all i and a, $\varphi_{f(e,i)}(a)$ is undefined, unless $a \in \{i-1, i+1\}$.

Exercises for Section 4.2

1. Discuss the validity of the s-m-n Theorem:

$$\varphi_{s_n^m(i,y_1,\ldots,y_m)}(z_1,\ldots,z_n) = \varphi_i(y_1,\ldots,y_m,z_1,\ldots,z_n)$$

   in the case where $\mathbf{P}_i$ has fewer than $m+n$ variables.

2. Show that there is a total computable function $g: \mathbf{N}^2 \to \mathbf{N}$ such that

$$\varphi_{g(i,j)}(x) = \varphi_i(x) + \varphi_j(x),$$

   where this sum is defined at x only when $\varphi_i(x)$ and $\varphi_j(x)$ are both defined.

3. Construct a total computable function f of two arguments such that $f(m,n)$ is an index for φ_m composed with itself n times.

4. For each of parts (a) and (b), prove the existence of a total computable function f having the specified property in two different ways—first, by explicitly defining the program-rewriting function f that transforms program x into program $f(x)$ and, second, by invoking the s-m-n Theorem and avoiding any program construction.

   (a) $\varphi_{f(x)}(y) = \varphi_x^{(2)}(y, \varphi_x^{(2)}(y,y))$,

   (b) $$\varphi_{f(x)}^{(2)}(y,z) = \begin{cases} \varphi_x^{(2)}(y,y), & \text{if } y = z; \\ \varphi_x^{(2)}(z,z), & \text{if } y = z+1; \\ \varphi_x^{(2)}(y,z), & \text{if } y = z+2; \\ \varphi_x^{(2)}(z,y), & \text{otherwise.} \end{cases}$$

5. Let $\psi_0, \psi_1, \ldots, \psi_n, \ldots$ be an arbitrary enumeration of computable functions, which may or may not be effective and which may or may not include all the computable functions; i.e., if we set $\psi_n = \varphi_{f(n)}$, the total function f is not necessarily computable nor necessarily onto. Let us call this enumeration $\{\psi_n \mid n \in \mathbf{N}\}$ *quasi-effective* if there is a total computable function $g: \mathbf{N} \to \mathbf{N}$ such that:

$$\varphi_{f(0)} = \varphi_{g(0)}, \varphi_{f(1)} = \varphi_{g(1)}, \ldots, \varphi_{f(n)} = \varphi_{g(n)}, \ldots .$$

Note that this does not necessarily mean that $g = f$. Prove that the enumeration $\{\psi_n \mid n \in \mathbf{N}\}$ is quasieffective if and only if it has a computable universal function $\Psi: \mathbf{N}^2 \to \mathbf{N}$. (Review the definition of "universal function" given in Exercise 6 of Section 3.2.)

4.3 Undecidable Problems

In computer science we often encounter families of *yes/no problems*, i.e., problems for which the answer is always either "yes" or "no". In relation to a family of yes/no problems, a computationally important question is to determine whether there exists an algorithm or *decision procedure* for solving all the problems in the family.

We say that a family of yes/no problems is *solvable* or *decidable* if there is an algorithm which, when presented with any problem in the family, will always converge with the correct "yes" or "no" answer. In our setting, the algorithm will be a **while**-program that, when given a numerical encoding of the problem as input, will output one of two values, 1 for "yes" and 0 for "no". If no such always-converging algorithm exists, we say that this family of yes/no problems is *unsolvable* or *undecidable*. We introduced the study of such problems in Section 1.2 when we showed that the Halting Problem was unsolvable. We showed that $halt: \mathbf{N} \to \mathbf{N}$ was *not* computable if

$$halt(i) = \begin{cases} 1, & \text{if } \varphi_i(i)\downarrow; \\ 0, & \text{otherwise.} \end{cases}$$

For another example, we may ask for each computable $\varphi_i: \mathbf{N} \to \mathbf{N}$ whether it is total or not, and then determine whether the collection of all such questions has a decision procedure. Put differently, is there one algorithm that will correctly answer any and every question from the family:

"Is $\varphi_i: \mathbf{N} \to \mathbf{N}$ a total function?", $i \in \mathbf{N}$?

We show below that this family of yes/no problems has no decision procedure, which we summarize by saying "it is unsolvable whether an arbitrary computable function φ_i is total or not."

Diagonalization and reduction

We often show that a family of yes/no problems is undecidable by assuming the opposite (i.e., that it is decidable) and deriving a contradiction from this assumption. There are two general methods of obtaining such a contradiction. The first method, called *diagonalization*, leads to a contradiction directly by applying an appropriately defined function to its own index. Our proof of the unsolvability of the Halting Problem given in Section 1.2 is an example of this method. The second technique, called *reduction*, establishes a contradiction indirectly, by showing that if the given family of yes/no problems were decidable, then it would contradict some already established result. (This "already established result" will usually be the undecidability of some other family of yes/no problems.) We met this method in Example 2 of Section 4.1 when we proved that ψ_2 was not computable by showing that its computability would yield a solution to the Halting Problem—surely a contradiction! Both diagonalization and reduction are powerful theoretical tools with wide applicability in mathematical logic and mathematics generally. We illustrate the method of diagonalization first to prove three basic facts about total functions, which are not themselves results about undecidability but which will be used later to establish such results.

1 Proposition. *There is a total function $f: \mathbf{N} \to \mathbf{N}$ which is not computable.*

Proof. Consider a countable but not necessarily effective enumeration of all the comp itable total functions from $\mathbf{N}$ to $\mathbf{N}$:

$$f_0, f_1, f_2, \ldots, f_i, \ldots, i \in \mathbf{N}.$$

For example, let $\varphi_0, \varphi_1, \varphi_2, \ldots$ be our standard enumeration of all computable functions. Let f_0 be the first total function in this sequence, and for each $n \geqslant 0$, let f_{n+1} be the first total φ_j occurring after f_n. This is certainly an enumeration of all total computable functions, although as we shall see in Corollary 3 it is *not* an effective enumeration.

Define $f: \mathbf{N} \to \mathbf{N}$ by

$$f(i) = f_i(i) + 1.$$

f is clearly a total function.

We show by contradiction that f cannot be computable. If f were computable, it would have an index j, say, in the above list of all total computable functions. This in turn would imply

$$f_j(j) = f_j(j) + 1,$$

which is clearly a contradiction. We are thus forced to conclude that f cannot be computable. □

Notice that the proof given above closely follows the Cantor diagonalization argument discussed in Section 1.3, if we think of a table with the sequence $f_i(0), f_i(1), f_i(2), \ldots, f_i(j), \ldots$ in row i, so that $f_i(i)$ is the "diagonal element" for each i.

For the next result, we first take another look at the notion of an "effective enumeration", introduced in Section 3.1. A detailed analysis will follow in Chapter 7. For now, we shall simply say that the sequence

$$\theta_0, \theta_1, \theta_2, \ldots, \theta_n, \ldots$$

of computable functions is an *effective enumeration* if there is some total computable function $g: \mathbf{N} \to \mathbf{N}$ such that $\theta_n = \varphi_{g(n)}$ for all $n \in \mathbf{N}$, where $\varphi_0, \varphi_1, \varphi_2, \ldots$ is our standard enumeration.

2 Proposition. *Let $f_0, f_1, \ldots, f_i, \ldots, i \in \mathbf{N}$, be an effective enumeration in which every f_i is a total computable function from* $\mathbf{N}$ *to* $\mathbf{N}$. *Then there is a total computable function $f: \mathbf{N} \to \mathbf{N}$ which does not appear in this effective enumeration.*

PROOF. To say that $f_0, f_1, \ldots, f_i, \ldots, i \in \mathbf{N}$, is an effective enumeration of total computable functions means that there is a total computable $g: \mathbf{N} \to \mathbf{N}$ such that $\varphi_{g(i)} = f_i$.

Define $f: \mathbf{N} \to \mathbf{N}$ by diagonalizing over the functions $f_0, f_1, \ldots, f_i, \ldots$; that is, let

$$f(i) = f_i(i) + 1.$$

f is a total function, and is computable by the **while**-program:

```
begin  X2 := g(X1);
       X1 := Φ(X2, X1);
       X1 := succ(X1)
end
```

But f cannot appear in the effective enumeration $f_0, f_1, \ldots, f_i, \ldots,$ itself, since, as in the previous proposition, it would differ from itself on its own index. □

3 Corollary. *There can be no effective enumeration of all the total computable functions from* $\mathbf{N}$ *to* $\mathbf{N}$.

Notice that the proposition and its corollary reinforce our claim in Chapter 1 that partial functions rather than total functions are the proper functional concepts for computability theory.

Using the *reduction* method, we next show that there can be no algorithm to decide whether an arbitrary computable function $\varphi_i: \mathbf{N} \to \mathbf{N}$ is total or not. Let us make explicit the "direction" in which this reduction goes: We prove that the "effective enumerability of the total computable functions" is *reducible* to the solvability of the Totality Problem, so that if the latter were solvable then the total computable functions would be effectively enumerable.

Our proof below is formal. An informal and considerably simpler proof based on Church's Thesis is left to the reader (Exercise 2).

4 Theorem. *There is no algorithm which, when presented with the index i of an arbitrary computable function $\varphi_i : \mathbf{N} \to \mathbf{N}$, can decide whether φ_i is total or not.*

PROOF. Define the total function *total*: $\mathbf{N} \to \mathbf{N}$ by

$$total(i) = \begin{cases} 1, & \text{if } \varphi_i : \mathbf{N} \to \mathbf{N} \text{ is total;} \\ 0, & \text{if not.} \end{cases}$$

Assume, by way of contradiction, that *total* is computable by some **while**-program. Then we may define a total computable function $g : \mathbf{N} \to \mathbf{N}$ by the scheme

$$g(0) = \text{the smallest } i \text{ such that } total(i) = 1,$$

$$g(n+1) = \text{the smallest } i \text{ greater than } g(n) \text{ such that } total(i) = 1.$$

In fact, g is computed by the following **while**-program, *if* we can assume that *total* is available as a subroutine. Given n as the initial data in $X1$, the program loads $n+1$ into $X2$, sets $X1$ to 0, and then computes until $X1$ holds the value $g(n)$.

```
begin X2 := succ(X1);
      X1 := 0;
      while X2 ≠ 0 do
           begin while total(X1) = 0 do X1 := succ(X1);
                X2 := pred(X2)
           end
end
```

Clearly,

$$\varphi_{g(0)}, \varphi_{g(1)}, \ldots, \varphi_{g(i)}, \ldots$$

is an enumeration containing all and only total computable functions. But this contradicts Corollary 3, and so *total* cannot be a computable function. □

The General Halting Problem for **while**-programs is also a family of yes/no problems, one for every pair consisting of a **while**-program P and an input vector $(a_1, \ldots, a_j)$ for P, stated as follows: "Does the computation by P on input $(a_1, \ldots, a_j)$ terminate?" We now prove that the General

Halting Problem, generalizing Section 1.2 and Theorem 1 of Section 3.1, is unsolvable. (With no loss of generality we consider only the case when **while**-programs are used to compute unary function, i.e., the case when $j = 1$.)

5 Theorem (Unsolvability of the General Halting Problem). *There is no algorithm which, when presented with the index i of an arbitrary computable function $\varphi_i : \mathbf{N} \to \mathbf{N}$ and an arbitrary argument a in $\mathbf{N}$, can decide whether $\varphi_i(a)$ is defined or not.*

PROOF. Define the total function

$$h(i,j) = \begin{cases} 1, & \text{if } \varphi_i(j) \text{ is defined;} \\ 0, & \text{otherwise.} \end{cases}$$

We want to prove that h cannot be computable. The proof is by reduction; that is, by reducing the computability of the function *halt* (defined in Theorem 1 of Section 3.1) to the computability of the function h. Indeed, $halt(i) = h(i,i)$ for all $i \in \mathbf{N}$. Thus, if h were computable, then *halt* would be too. But this contradicts Theorem 1 of Section 3.1, which forces us to conclude that h cannot be computable either. □

The Equivalence Problem for **while**-programs is a family of yes/no problems, one for every pair of **while**-programs P and P', stated as follows: "Do P and P' compute the same function?"

We prove the unsolvability of the Equivalence Problem by two successive reductions. The result established by the first reduction is itself an unsolvability result: It is undecidable whether a given **while**-program computes a specified function. This is the important Verification Problem of computer science. Lemma 6 shows that this problem is unsolvable even for the simple case where the function specified is the identify function, $id(x) = x$ for all x.

6 Lemma. *Consider the family of all computable functions $\varphi_i : \mathbf{N} \to \mathbf{N}$, $i \in \mathbf{N}$. The function*

$$e(i) = \begin{cases} 1, & \textit{if } \varphi_i = id, \\ 0, & \textit{otherwise,} \end{cases}$$

is not effectively computable.

PROOF. We first define the program

$$\mathbf{P}_{g(i)} = \textbf{begin } X2 := \Phi(i, X1) \textbf{ end}$$

where g is a total computable function by the s-m-n Theorem. Hence,

$$\varphi_{g(i)}(j) = \begin{cases} j, & \text{if } \varphi_i(j)\downarrow; \\ \perp, & \text{otherwise.} \end{cases}$$

Observe now that $\varphi_{g(i)}$ is the identity function *id* just in case φ_i is a total function:

$$\varphi_{g(i)} = id \Leftrightarrow \varphi_i \text{ is total.}$$

Consider next the composition of the function e, defined in the statement of the lemma, and g:

$$e(g(i)) = \begin{cases} 1, & \text{if } \varphi_{g(i)} = id; \\ 0, & \text{otherwise.} \end{cases}$$

By our last observation,

$$e(g(i)) = \begin{cases} 1, & \text{if } \varphi_i \text{ is total;} \\ 0, & \text{otherwise.} \end{cases}$$

This means that the function *total*, defined in Theorem 4, is none other than $e \circ g$. We have thus reduced the computability of *total* to that of e. Since we know that *total* is not computable, we conclude that e cannot be either. □

Many generalizations of the above proposition are possible. We may for example replace the identity function *id* by any other specified function, which must, however, be total and computable in order that (a slightly amended version of) the proof still apply. We may also generalize the proposition to the family of all computable functions $\varphi_i^{(j)} : \mathbf{N}^j \rightarrow \mathbf{N}$, $i \in \mathbf{N}$, for arbitrary $j \geqslant 1$, and the proof technique still applies.

7 Theorem (Unsolvability of the Equivalence Problem). *There is no algorithm which, when presented with indices i and j of arbitrary computable functions* $\varphi_i : \mathbf{N} \rightarrow \mathbf{N}$ *and* $\varphi_j : \mathbf{N} \rightarrow \mathbf{N}$ *can decide whether* $\varphi_i = \varphi_j$.

PROOF. Define the total function

$$equivalent(i, j) = \begin{cases} 1, & \text{if } \varphi_i = \varphi_j; \\ 0, & \text{otherwise.} \end{cases}$$

The identity function is certainly computable; let j be an index for it, so that $id = \varphi_j$, e.g., $\mathbf{P}_j =$ **begin end**. It should be clear that we can now write $e(i) = equivalent(i, j)$ for all $i \in \mathbf{N}$, where $e : \mathbf{N} \rightarrow \mathbf{N}$ was defined in the preceding lemma. We have thus reduced the computability of e to that of *equivalent*. Since e is not computable, *equivalent* cannot be either. □

EXERCISES FOR SECTION 4.3

1. Review the definition of "computable real numbers" given in Exercise 6 of Section 1.3 (with "PASCAL program" replaced by "**while**-program"). Prove that the computable real numbers cannot be effectively enumerated. With no loss of generality (why?), you can restrict the computable reals to those lying between 0 and 1.

2. Use Church's Thesis to give an informal but rigorous argument that if a set of numbers cannot be effectively generated, then it is undecidable whether an arbitrary number is in the set.

3. Prove that there is no total computable function $f: \mathbf{N}^2 \to \mathbf{N}$ such that for all i and j, $\mathbf{P}_i$ on input j halts $\Leftrightarrow \mathbf{P}_i$ on input j halts within $f(i, j)$ steps. (*Hint*: The existence of f would solve the General Halting Problem.)

4. (a) Reduce the solvability of the set $\{i \mid \varphi_i(i)\downarrow\}$ to the solvability of

$$\{(i, j) \mid \varphi_i(\varphi_j(i))\downarrow\}$$

to show that the latter is not decidable.

(b) By a technique of your own choosing, prove that the set $\{i \mid \varphi_i(i^2 + i + 1)\downarrow\}$ is unsolvable.

5. Redo the proof of Lemma 6 to establish the unsolvability of the set $\{n \mid \varphi_n \equiv 0\}$; that is, it is impossible to verify algorithmically that an arbitrary **while**-program computes the function which identically equals 0.

6. The definition of function g in the proof of Lemma 6 is an application of the s-m-n Theorem (program form). Redefine g by invoking Church's Thesis and then explicitly the s-m-n Theorem (functional form), avoiding any program construction.

7. Using the function g defined in the proof of Lemma 6, prove that none of the functions in (1), (2), (3), (4), and (5) below are computable.

(1) "It is undecidable whether an arbitrary **while**-program outputs infinitely many values."

$$f_1(i) = \begin{cases} 1, & \text{if RAN } (\varphi_i) \text{ is infinite;} \\ 0, & \text{otherwise.} \end{cases}$$

(2) "It is undecidable whether an arbitrary **while**-program halts on some input value."

$$f_2(i) = \begin{cases} 1, & \text{if DOM } (\varphi_i) \neq \emptyset; \\ 0, & \text{otherwise.} \end{cases}$$

(3) "It is undecidable whether an arbitrary **while**-program halts on infinitely many input values."

$$f_3(i) = \begin{cases} 1, & \text{if DOM } (\varphi_i) \text{ is infinite;} \\ 0, & \text{otherwise.} \end{cases}$$

(4) "It is undecidable whether an arbitrary **while**-program halts on all input values" (alternative proof of the unsolvability of the Totality Problem).

$$f_4(i) = \begin{cases} 1, & \text{if DOM } (\varphi_i) = \mathbf{N}; \\ 0, & \text{otherwise.} \end{cases}$$

(5)

$$f_5(i, j, k) = \begin{cases} 1, & \text{if } \varphi_i(j) = k; \\ 0, & \text{otherwise.} \end{cases}$$

(*Hint*: The computability of $f_1 \circ g$, $f_2 \circ g$, $f_3 \circ g$, or $f_4 \circ g$ implies the solvability of the Halting Problem. The computability of $f_5(g(x), 0, 0)$ as a function of x implies the same.)

8. (a) Prove the existence of a total computable function g such that for all i and j:

$$\varphi_{g(i)}(j) = \begin{cases} \varphi_j(j) + 1, & \text{if } \varphi_i(i)\downarrow; \\ \perp, & \text{otherwise.} \end{cases}$$

Show that for all $i, \varphi_{g(i)}$ can be extended to a total computable function $\Leftrightarrow \varphi_{g(i)}$ identically equal to the empty function.

(b) Use the function g of part (a) to prove that it is undecidable whether an arbitrary computable function is extendible to a total computable function.

9. Consider the total function $f: \mathbf{N} \to \mathbf{N}$ which maps $\mathbf{P}_n$ to the "shortest" **while**-program computing the same unary function as $\mathbf{P}_n$; that is, for all $n \in \mathbf{N}$, $f(n)$ is the smallest index such that $\varphi_n = \varphi_{f(n)}$. Prove that f cannot be effective. (*Hint*: The computability of f would contradict the unsolvability of the Equivalence Problem.)

10. Consider the following total function $F: \mathbf{N} \to \mathbf{N}$:

$F(n) =$ "largest possible output (i.e., largest value stored in variable $X1$ at exit) if and when a **while**-program, with exactly n basic instructions, halts on input 0."

Recall that a basic instruction is of the form $X := 0$ or $X := succ(Y)$ or $X := pred(Y)$ or $X \neq Y$? If a **while**-program P has exactly n basic instructions, the flow diagram of P contains n "boxes" (assignments) and "diamonds" (tests).

(a) Show that F is a well-defined total function; that is, show that there are finitely many **while**-programs with exactly n basic instructions, and that there is a uniform (least) upper bound $F(n)$ on the outputs of all such **while**-programs that halt on input 0. Compute the values of $F(0)$, $F(1)$, $F(2)$, $F(3)$, and $F(4)$. Since a basic instruction uses at most two distinct variables, assume with no loss of generality (why?) that the variables of a **while**-program with n basic instructions are all in the set $\{X1, \ldots, X2n\}$.

We want to show that F is not computable, by a new form of diagonalization. To this end, we say that the total function $f: \mathbf{N} \to \mathbf{N}$ is *bounded* by a total function $g: \mathbf{N} \to \mathbf{N}$ if $f(x) \leqslant g(x)$ for all x. We say f is *majorized* by g if there is a number k such that for all $x > k$, $f(x) \leqslant g(x)$.

(b) Show that every total computable function $f: \mathbf{N} \to \mathbf{N}$ is bounded by a strictly increasing total computable function $g: \mathbf{N} \to \mathbf{N}$, which also satisfies the condition $g(x) > x^2$ for all x.

(c) Prove that every strictly increasing total computable function $g: \mathbf{N} \to \mathbf{N}$ such that $g(x) > x^2$ for all x is majorized by the function F. (*Hint*: If $g = \varphi_e$ and $size(e)$ is the number of basic instructions in $\mathbf{P}_e$, consider **while**-program

$$\textbf{begin } X1 := n;\ X1 := g(X1);\ X1 := g(X1) \textbf{ end}$$

to conclude that $F(n + 2\, size(e)) \geqslant g(g(n))$ for all $n \in \mathbf{N}$. Then using the fact that $g(n) > n^2 > n + 2\, size(e)$ for sufficiently large n, conclude that F majorizes g.)

(d) Use (b) and (c) to conclude that F majorizes every total computable function, and is therefore not computable.

11. This continues the discussion of the preceding exercise. We want to prove that the following step-counting function is not effective:

$G(n) =$ "maximum number of steps before a **while**-program, with exactly n basic instructions, ever halts on input 0."

(a) Show that G is a well-defined total function. Compute the values of $G(0)$, $G(1)$, $G(2)$, $G(3)$, and $G(4)$.

(b) Prove that G is not computable—in two different ways:

(i) by using the fact that F of the preceding exercise is not effective, and showing that F is bounded by G;

(ii) by showing that if G were computable then there would be a decision procedure for membership in the set $\{n \mid \varphi_n(0)\downarrow\}$.

CHAPTER 5

Program Methodology

We have now introduced the language of **while**-programs and shown how, with the use of macros, we can attain much of the power of a "real" programming language like PASCAL. We have gone far to justify the claim that **while**-programs are complete, i.e., that any algorithm at all can be translated into a **while**-program. As we have already seen in our discussion of various undecidable problems in Section 4.3, this very breadth of expressive power exacts a price. Many interesting questions about programs cannot be answered by an algorithm which is guaranteed to work for *all* programs as input data.

This chapter, then, provides something of an "antidote". We show that there is a methodology which allows us to prove properties of many of the programs encountered in normal programming practice. As we have already seen, computability theory had its roots in the 1930's and 1940's in the work of people like Church, Gödel, Kleene, Turing, and Post. Stored program computers were an invention of the 1940's, and the development of ALGOL-like languages—of which our language of **while**-programs is an example—did not occur until the late 1950's. While the logicians of the 1930's and 1940's concentrated on problems which were unsolvable, computer scientists from the 1960's on have developed a *programming methodology* which allows us, in many cases, to formally prove that a program does indeed solve the problem it was intended to solve.

The semantics of a program is the function it computes. Our approach to program semantics in Section 2.3 was *operational*, because it involved tracing the sequence of operations that each program execution gave rise to. In Section 5.1 we introduce the *denotational* approach to program semantics. In this approach we map all syntactic constructs in a program to the abstract values (i.e., numbers and functions) which they "denote". The resulting semantics of programs is based on their inductive structure. The value denoted by a syntactic construct is determined by the values denoted by its constituent parts. Programs can then be interpreted as functions from initial states to final states without mention of intermediate states.

By defining program meanings abstractly, we also show that denotational semantics provides us with elegant tools for proving equivalence of syntactic constructs. We shall study two approaches to denotational semantics, the order semantics of Scott and Strachey (1971) and the partially additive semantics of Arbib and Manes (1982).

In Section 5.2 we prove that functions computed by "recursive programs" are already computed by **while**-programs. We then use techniques introduced in Section 5.1 to establish the First Recursion Theorem. This discussion will provide valuable motivation for the discussion of the (Second) Recursion Theorem in Section 6.1.

In Section 5.3 we provide a different program methodology, which we call *logical semantics*. The results of this section will not be used elsewhere in the book, but are included since they provide a widely used set of techniques for proving program correctness. In *logical semantics*, we do not study transformations of states directly, but rather look at logical predicates (which can be asserted) of the state of computation. We present the proof rule methodology of Floyd (1967) and Hoare (1969). Their approach was extended to a semantic definition of most of PASCAL by Hoare and Wirth (1973).

5.1 An Invitation to Denotational Semantics

In Section 2.3 we set up conventions for associating computable functions with **while**-programs. For convenience we reformulate these conventions in the following way:

(1) With every k-variable **while**-program P, we associate a function $\alpha_P : \mathbf{N}^k \to \mathbf{N}^k$ defined by

$$\alpha_P(a_1, \ldots, a_k) = \begin{cases} (b_1, \ldots, b_k), & \text{if the computation of } P \text{ on input vector } (a_1, \ldots, a_k) \text{ halts, and the final state vector is } (b_1, \ldots, b_k); \\ \perp, & \text{otherwise.} \end{cases}$$

Recall that we require that $k \geqslant 1$ by our conventions in Section 2.3.

(2) For every $j \geqslant 1$ and $k \geqslant 1$, we define an input map $in : \mathbf{N}^j \to \mathbf{N}^k$ which sends $(a_1, \ldots, a_j)$ to $(a_1, \ldots, a_j, 0, \ldots, 0)$ if $j < k$, and to $(a_1, \ldots, a_k)$ if $j \geqslant k$. We also define an output map $out : \mathbf{N}^k \to \mathbf{N}$ which sends $(a_1, \ldots, a_k)$ to a_1.

For any particular $j \geqslant 1$ and any k-variable **while**-program P, we clearly have $\varphi_P^{(j)} = out \circ \alpha_P \circ in$. When convenient, we shall consider α_P instead of $\varphi_P^{(j)}$.

The definition of α_P given in (1) is based on the notion of a "computation sequence", as introduced in Section 2.3. Since a computation sequence is any path through the flow diagram of P satisfying certain consistency conditions, P need not be a strict **while**-program for the concept of a computation sequence to make sense (for example, P may be a flowchart or a **goto**-program as defined in Exercises 2 and 3 of Section 2.1). Our definition of operational semantics is therefore more general than needed for a discussion restricted to **while**-programs.

We now exploit the inductive structure of **while**-programs to give an equivalent definition of their semantics. This inductive definition will make a comparison with denotational semantics easier, since denotational semantics is also based on the inductive structure of **while**-programs. We call this first inductive definition the *basic* semantics.

1 Definition (Basic Semantics of k-Variable **while**-Programs)

(1) If $\mathbb{S}$ is the assignment statement $Xu := g(Xv)$ where $g(Xv)$ is either $succ(Xv)$ or $pred(Xv)$ or 0, then $\alpha_{\mathbb{S}} : \mathbf{N}^k \to \mathbf{N}^k$ is the function which sends $(a_1, \ldots, a_k) \in \mathbf{N}^k$ to $(a_1, \ldots, a_{u-1}, g(a_v), a_{u+1}, \ldots, a_k)$.

(2) If $\mathbb{S}$ is the compound statement **begin** $\mathbb{S}_1; \ldots; \mathbb{S}_m$ **end** where $\alpha_{\mathbb{S}_i} : \mathbf{N}^k \to \mathbf{N}^k$ is the semantics of $\mathbb{S}_i$, for $1 \leqslant i \leqslant m$, then we define $\alpha_{\mathbb{S}} : \mathbf{N}^k \to \mathbf{N}^k$ by

$$\alpha_{\mathbb{S}} = \alpha_{\mathbb{S}_m} \circ \cdots \circ \alpha_{\mathbb{S}_2} \circ \alpha_{\mathbb{S}_1}$$

(3) If $\hat{\mathbb{S}}$ is the **while** statement

$$\textbf{while } Xu \neq Xv \textbf{ do } \mathbb{S}$$

then $\alpha_{\hat{\mathbb{S}}} : \mathbf{N}^k \to \mathbf{N}^k$ is defined by

$$\alpha_{\hat{\mathbb{S}}}(\mathbf{a}) = \begin{cases} \alpha_{\mathbb{S}}^n(\mathbf{a}), & \text{if there is an } n \geqslant 0 \text{ such that} \\ & \left[\alpha_{\mathbb{S}}^n(\mathbf{a})\right]_u = \left[\alpha_{\mathbb{S}}^n(\mathbf{a})\right]_v \text{ and for all} \\ & 0 \leqslant m \leqslant n-1, \left[\alpha_{\mathbb{S}}^m(\mathbf{a})\right]_u \neq \left[\alpha_{\mathbb{S}}^m(\mathbf{a})\right]_v; \\ \perp, & \text{otherwise;} \end{cases}$$

where $\mathbf{a} = (a_1, \ldots, a_k)$, and $[\mathbf{x}]_u$ denotes the uth component of the k-dimensional vector $\mathbf{x}$.

We leave it to the reader to verify that for any k-variable **while**-program P, its basic semantics α_P as defined above is identical to that obtained by tracing paths through the flow diagram of P (Exercise 1).

We now turn to the denotational semantics of **while**-programs. We study two approaches to denotational semantics, the *partially additive semantics* of Arbib and Manes (1982) and the *order semantics* of Scott and Strachey (1971). Both approaches are based on the inductive structure of the syntax, and both exploit algebraic properties of the functions computed by **while**-programs.

Partially additive semantics

To motivate this approach, consider the conditional statement $\mathbb{S}$ = **if** $\mathcal{C}$ **then** $\mathbb{S}_1$ **else** $\mathbb{S}_2$ with the following flow diagram.

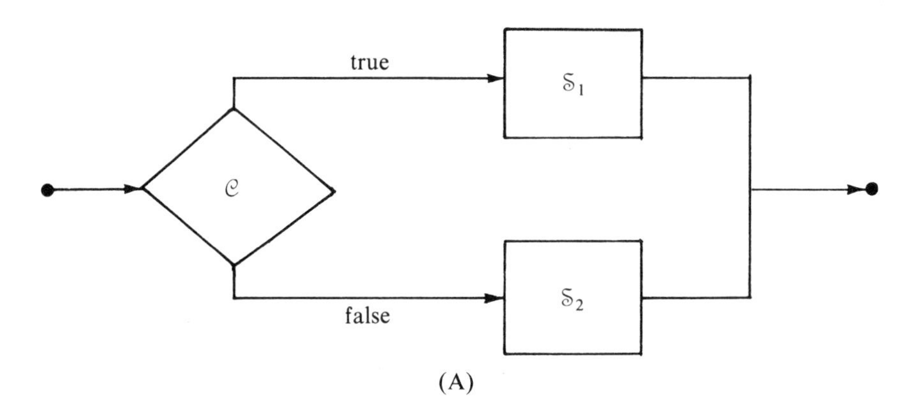

(A)

Let the test $\mathcal{C}$ be interpreted by predicate $p:\mathbf{N}^k \to \{\mathit{true},\ \mathit{false}\}$. For example, if $\mathcal{C}$ is the test $X2 \neq X3$ we have

$$p(a_1, a_2, a_3, \ldots, a_k) = \begin{cases} \mathit{true}, & \text{if } a_2 \neq a_3\,; \\ \mathit{false}, & \text{if } a_2 = a_3\,. \end{cases}$$

Assuming that $\mathbb{S}_1$ and $\mathbb{S}_2$ have interpretations $\beta_1 : \mathbf{N}^k \to \mathbf{N}^k$ and $\beta_2 : \mathbf{N}^k \to \mathbf{N}^k$, respectively, the interpretation of the conditional $\mathbb{S}$ is given by

$$\beta_{\mathbb{S}}(\mathbf{a}) = \begin{cases} \beta_1(\mathbf{a}), & \text{if } p(\mathbf{a}) = \mathit{true}; \\ \beta_2(\mathbf{a}), & \text{if } p(\mathbf{a}) = \mathit{false}. \end{cases}$$

(We use β's instead of α's to distinguish partially additive semantics from basic semantics. Later we shall see that for all **while**-programs P, $\alpha_P = \beta_P$.)

We now analyze the conditional $\mathbb{S}$ using partially additive semantics. Note that there are two paths from entry to exit in diagram (A). We associate a semantics with each separate path by first transforming (A) to the following form.

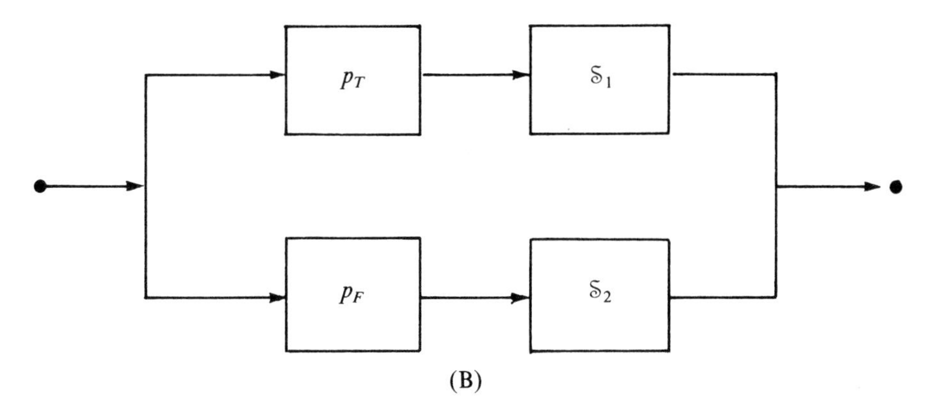

(B)

Here we have replaced the test with interpretation p by a line splitting to feed two partial functions:

$$p_T(\mathbf{a}) = \begin{cases} \mathbf{a}, & \text{if } p(\mathbf{a}) = \mathit{true}; \\ \bot, & \text{otherwise.} \end{cases}$$

$$p_F(\mathbf{a}) = \begin{cases} \mathbf{a}, & \text{if } p(\mathbf{a}) = \mathit{false}; \\ \bot, & \text{otherwise.} \end{cases}$$

We then note that both (A) and (B) share the property that the upper path has as semantics the composition $\beta_1 \circ p_T$, i.e., it is only taken if $p(\mathbf{a}) = \mathit{true}$ and then transforms $\mathbf{a}$ to $\beta_1(\mathbf{a})$, while the lower path has semantics $\beta_2 \circ p_F$. Noting that the two partial functions $\beta_1 \circ p_T$ and $\beta_2 \circ p_F$ are disjoint (that is, their domains of definitions are disjoint), we may define their "sum" $\beta_1 \circ p_T + \beta_2 \circ p_F$ by

$$(\beta_1 \circ p_T + \beta_2 \circ p_F)(\mathbf{a}) = \begin{cases} \beta_1 \circ p_T(\mathbf{a}), & \text{if } \mathbf{a} \in \mathrm{DOM}(\beta_1 \circ p_T); \\ \beta_2 \circ p_F(\mathbf{a}), & \text{if } \mathbf{a} \in \mathrm{DOM}(\beta_2 \circ p_F). \end{cases}$$

We conclude that the conditional $\mathbb{S}$ has the interpretation

$$\beta_{\mathbb{S}} = \beta_1 \circ p_T + \beta_2 \circ p_F.$$

This motivates the following definition.

2 Definition. For any two sets A and B, let $\mathbf{Pfn}(A, B)$ denote the set of all partial functions from A to B. We say a family $\{\theta_i \mid i \in I\}$ of functions in $\mathbf{Pfn}(A, B)$ is *disjoint* if their domains of definition are disjoint. We define the *sum* $\Sigma_{i \in I} \theta_i$ (or simply $\Sigma \theta_i$ if the index set I is understood) of the disjoint family $\{\theta_i\}$ by

$$\left(\sum \theta_i\right)(x) = \begin{cases} \theta_i(x), & \text{if } x \in \mathrm{DOM}(\theta_i) \text{ for some } i \in I; \\ \bot, & \text{if no such } i \text{ exists.} \end{cases}$$

The sum is well defined by disjointness of the θ's. Σ is a *partial* addition since it is defined only for disjoint families or θ's.

3 Proposition. *Let ζ be in* **Pfn**(A, B), $\{\theta_i \mid i \in I\}$ *be a disjoint family in* **Pfn**(B, C), *and ξ be in* **Pfn**(C, D). *Then the family* $\{\xi \circ \theta_i \circ \zeta \mid i \in I\}$ *is disjoint in* **Pfn**(A, D) *and*

$$\sum \xi \circ \theta_i \circ \zeta = \xi \circ \left(\sum \theta_i\right) \circ \zeta.$$

PROOF. Exercise 2. □

We now consider the partially-additive semantics of iterative statements **while-do** and **repeat-until**. In Figures 1(a) and 1(b) we have redrawn the **while** $\mathcal{C}$ **do** $\mathcal{S}$ and **repeat** $\mathcal{S}$ **until** $\mathcal{C}$ constructs so that we view them as special cases of the iteration construct $K^\dagger$ (read "K dagger") shown in Figure 1(c). The *iterate* $K^\dagger$ is obtained by connecting one of K's exit points to its entry point.

Assume that the one-entry, two-exit flow diagram K is given the following interpretation:

$$\beta : \mathbf{N}^k \to \mathbf{N}^k + \mathbf{N}^k$$

where $\mathbf{N}^k + \mathbf{N}^k = \mathbf{N}^k \times \{1\} \cup \mathbf{N}^k \times \{2\}$ is the disjoint union of two copies of $\mathbf{N}^k$. If we write $\beta(\mathbf{a}) = (\mathbf{b}, i)$ with $i \in \{1,2\}$, we mean that K on input $\mathbf{a}$ yields output $\mathbf{b}$ on line i. As shown in Figure 2(a), we associate with β partial functions β_1 and β_2 mapping $\mathbf{N}^k$ to $\mathbf{N}^k$, corresponding to which exit is taken:

$$\beta_i(\mathbf{a}) = \begin{cases} \mathbf{b}, & \text{if } \beta(\mathbf{a}) = (\mathbf{b}, i); \\ \bot, & \text{otherwise.} \end{cases}$$

β_1 and β_2 are clearly disjoint.

If β is the interpretation of K, then we may form the interpretation $\beta^\dagger$ of $K^\dagger$ by the informal rule "evaluate β; if exit is on line 1, repeat; otherwise halt". Given input $\mathbf{a}$, either there is some exact number $n \geqslant 0$ times around the loop prior to exit, in which case [see Figure 2(b)] $\beta^\dagger(\mathbf{a}) = \beta_2 \circ \beta_1^n(\mathbf{a})$; or else the computation gets "caught in the loop", in which case $\beta^\dagger(\mathbf{a})$ is undefined. Hence we can write

$$\beta^\dagger(\mathbf{a}) = \begin{cases} \beta_2 \circ \beta_1^n(\mathbf{a}), & \text{if there is an } n \geqslant 0 \text{ such that} \\ & \beta_2 \circ \beta_1^n(\mathbf{a}) \text{ is defined and for all} \\ & 0 \leqslant m \leqslant n-1,\ \beta_2 \circ \beta_1^m(\mathbf{a}) = \bot; \\ \bot, & \text{otherwise.} \end{cases}$$

The following result is crucial for later development. It will allow us to put $\beta^\dagger$ in a more convenient form.

4 Lemma. *Let $\zeta, \xi \in$* **Pfn**(A, A). *If ζ and ξ are disjoint, then so are $\zeta \circ \xi^m$ and $\zeta \circ \xi^n$ for all $m \neq n$.*

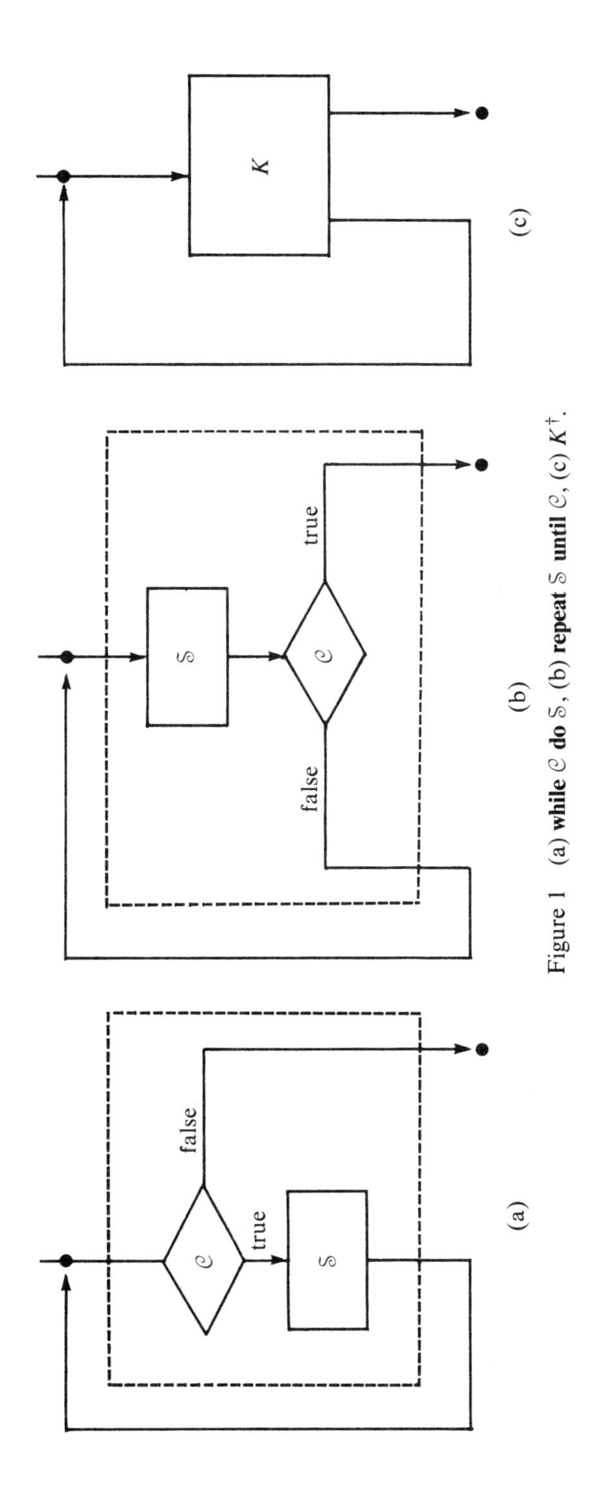

Figure 1 (a) **while** $\mathcal{C}$ **do** $\mathcal{S}$, (b) **repeat** $\mathcal{S}$ **until** $\mathcal{C}$, (c) $K^{\dagger}$.

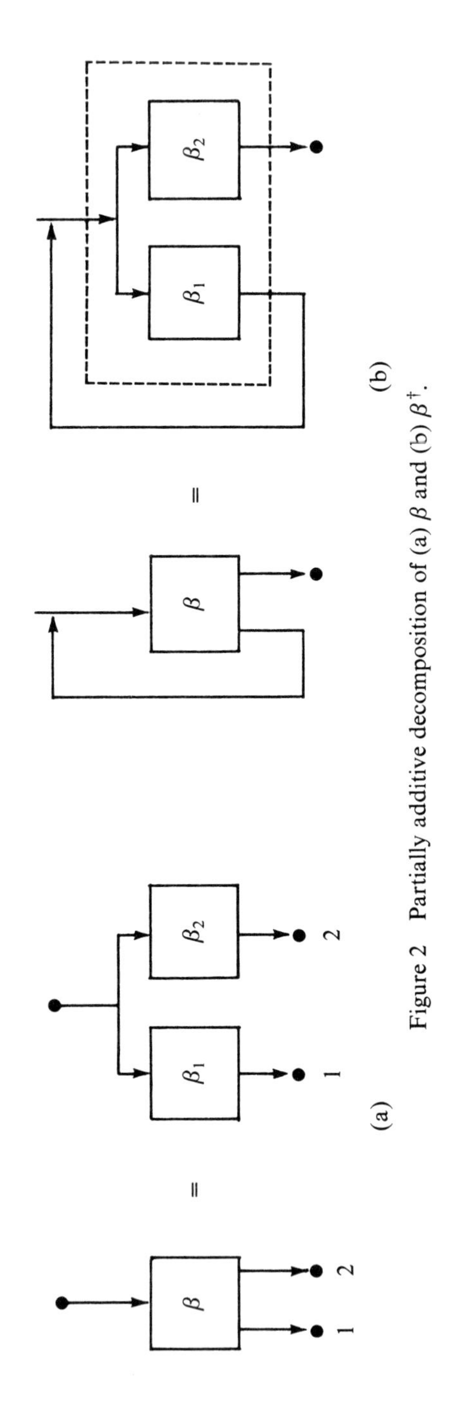

Figure 2 Partially additive decomposition of (a) β and (b) $\beta^{\dagger}$.

PROOF. Assume $m < n$, and consider $\zeta \circ \xi^m$ and $\xi \circ \xi^m$. Since $\mathrm{DOM}(\zeta) \cap \mathrm{DOM}(\xi) = \emptyset$, it is easily checked that $\mathrm{DOM}(\zeta \circ \xi^m) \cap \mathrm{DOM}(\xi \circ \xi^m) = \emptyset$. Hence $\mathrm{DOM}(\zeta \circ \xi^m) \cap \mathrm{DOM}(\theta \circ \xi \circ \xi^m) = \emptyset$ for any $\theta \in \mathbf{Pfn}(A, A)$. In particular, $\mathrm{DOM}(\zeta \circ \xi^m) \cap \mathrm{DOM}(\zeta \circ \xi^n) = \emptyset$, as desired. □

5 Lemma. *Let the one-entry, two-exit flow diagram K have interpretation $\beta : \mathbf{N}^k \to \mathbf{N}^k + \mathbf{N}^k$ inducing the disjoint functions $\beta_1, \beta_2 \in \mathbf{Pfn}(\mathbf{N}^k, \mathbf{N}^k)$. Then the interpretation $\beta^\dagger$ of the iterate $K^\dagger$ is*

$$\beta^\dagger = \sum_{n \geq 0} \beta_2 \circ \beta_1^n .$$

PROOF. By Lemma 4, the family $\{ \beta_2 \circ \beta_1^n \mid n \geq 0\}$ of partial functions is disjoint. Their sum is therefore well defined. □

For the purpose of comparison with operational semantics, we need the following definition.

6 Definition (Partially Additive Semantics of **while**-Programs). We assign to every k-variable **while**-program P an interpretation $\beta_P : \mathbf{N}^k \to \mathbf{N}^k$ defined inductively by:

(1) Identical to (1) of Definition 1.
(2) Identical to (2) of Definition 1.

[In both (1) and (2) we substitute β for α throughout.]

(3) If $\hat{\mathbb{S}}$ is the **while** statement

$$\textbf{while } Xu \neq Xv \textbf{ do } \mathbb{S}$$

then $\beta_{\hat{\mathbb{S}}} = \Sigma_{n \geq 0}\, p_F \circ (\beta_{\mathbb{S}} \circ p_T)^n$, where

$$p_T(a_1, \ldots, a_k) = \begin{cases} (a_1, \ldots, a_k), & \text{if } a_u \neq a_v\,; \\ \bot, & \text{if } a_u = a_v\,; \end{cases}$$

$$p_F(a_1, \ldots, a_k) = \begin{cases} (a_1, \ldots, a_k), & \text{if } a_u = a_v\,; \\ \bot, & \text{if } a_u \neq a_v\,. \end{cases}$$

The justification for part (3) in the above definition follows from Lemma 5, where we substitute $\beta_{\mathbb{S}} \circ p_T$ for β_1 and p_F for β_2 (see Figures 1(a) and 2(b)).

7 Theorem. *For every **while**-program P, $\alpha_P = \beta_P$; that is, the basic semantics of **while**-programs are equal to their partially additive semantics.*

PROOF. By induction on the structure of **while**-programs, we prove for every statement $\mathbb{S}$ that $\alpha_{\mathbb{S}} = \beta_{\mathbb{S}}$.

If $\mathbb{S}$ is an assignment or a compound statement, then trivially $\alpha_{\mathbb{S}} = \beta_{\mathbb{S}}$, since parts (1) and (2) of Definitions 1 and 6 are identical.

If $\hat{\mathbb{S}}$ is the **while** statement **while** $Xu \neq Xv$ **do** $\mathbb{S}$, we assume that $\alpha_{\mathbb{S}} = \beta_{\mathbb{S}}$ (induction hypothesis) and prove that $\alpha_{\hat{\mathbb{S}}} = \beta_{\hat{\mathbb{S}}}$. This is easily established using the definition and Lemma 5. Details are left to the reader (Exercise 3). □

Partially additive semantics provides us with convenient tools for proving the equivalence of different program constructs. A simple example follows. Others appear in the exercises.

8 Example. Let $\mathcal{C}$ have interpretation $p:\mathbf{N}^k \to \{true, false\}$, and $\mathbb{S}$ have interpretation $\beta:\mathbf{N}^k \to \mathbf{N}^k$. The interpretation of **while** $\mathcal{C}$ **do** $\mathbb{S}$ is

$$\beta_{\textbf{while } \mathcal{C} \textbf{ do } \mathbb{S}} = \sum_{n \geqslant 0} p_F \circ (\beta \circ p_T)^n,$$

by Lemma 5, where we substitute $\beta \circ p_T$ for β_1 and p_F for β_2 [see also Figure 1(a)]. In the same way, the interpretation of **repeat** $\mathbb{S}$ **until** $\mathcal{C}$ is

$$\beta_{\textbf{repeat } \mathbb{S} \textbf{ until } \mathcal{C}} = \sum_{n \geqslant 0} (p_T \circ \beta) \circ (p_F \circ \beta)^n$$

by Lemma 5 [see Figure 1(b)]. Let us consider the latter equation further:

$$\begin{aligned}\beta_{\textbf{repeat } \mathbb{S} \textbf{ until } \mathcal{C}} &= \sum (p_T \circ \beta) \circ (p_F \circ \beta)^n \\ &= \sum p_T \circ (\beta \circ p_F)^n \circ \beta \\ &= \left[\sum (\neg p)_F \circ (\beta \circ (\neg p)_T)^n \right] \circ \beta,\end{aligned}$$

by Proposition 3, and noting $(\neg p)_T = p_F$ and $(\neg p)_F = p_T$:

$$\begin{aligned}&= \left[\beta_{\textbf{while } \neg \mathcal{C} \textbf{ do } \mathbb{S}} \right] \circ \beta \\ &= \beta_{\textbf{begin } \mathbb{S}; \textbf{ while } \neg \mathcal{C} \textbf{ do } \mathbb{S} \textbf{ end}}\end{aligned}$$

For further discussion and applications of partially additive semantics, the reader is referred to Arbib and Manes (1982).

Order semantics

Partially additive semantics exploits the partial addition Σ defined on $\mathbf{Pfn}(\mathbf{N}^k, \mathbf{N}^k)$ to define the iterate $\beta^\dagger$ via a "sum of paths", with one term $\beta_2 \circ \beta_1^n$ for each *exact* number n of times around the loop. By contrast, *order semantics* exploits the existence of a partial order $\leqslant$ on $\mathbf{Pfn}(\mathbf{N}^k, \mathbf{N}^k)$:

$$\zeta \leqslant \xi \text{ iff } \xi \text{ extends } \zeta\text{, that is, } \mathrm{DOM}(\zeta) \subset \mathrm{DOM}(\xi),$$
$$\text{with } \xi(\mathbf{a}) = \zeta(\mathbf{a}) \text{ for each } \mathbf{a} \text{ in } \mathrm{DOM}(\zeta).$$

It then notes that terms $\Sigma_{j \leqslant n}\, \beta_2 \circ \beta_1^j$ form an increasing sequence, where the nth term corresponds to *at most n* times around the loop. The semantics of the loop construction is obtained by passing to the limit of the sequence of bounded sums. We now formalize this in a more general setting.

9 Definition. An *ω-complete partially ordered set* (abbreviated *ω-cpo*) is a poset[1] $(A, \leqslant)$ equipped with a *minimal element* $\perp$ (read "bottom"),

$$\perp \leqslant a \text{ for all } a \text{ in } A$$

which is *ω-complete* in the sense that any ω-chain

$$a_0 \leqslant a_1 \leqslant a_2 \leqslant \cdots \leqslant a_n \leqslant \cdots$$

has a *least upper bound* (abbreviated *lub*)

$$a = \bigvee_{n \geqslant 0} a_n ,$$

i.e., $a_n \leqslant a$ for each n, while any $\bar{a}$ with $a_n \leqslant \bar{a}$ for any n must satisfy $a \leqslant \bar{a}$.

10 Examples. (Example (d) is the crucial one for our semantics.)

(a) The poset $(2^S, \subset)$ of subsets of a set S, ordered under set inclusion, is an ω-cpo with minimal element $\emptyset$, the empty set, while $\bigvee_{n \geqslant 0} a_n = \bigcup_{n \geqslant 0} a_n$, the union of the sets of the ω-chain.
(b) The poset $(\mathbf{N}, \leqslant)$ of natural numbers under the usual less-than-or-equal-to-ordering is not an ω-cpo. It has minimal element 0, but an ω-chain (a_n) has an upper bound only if $max\{a_n \mid n \geqslant 0\}$ is finite.
(c) However, if we extend $\leqslant$ from $\mathbf{N}$ to $\mathbf{N} \cup \{\infty\}$ by setting $n \leqslant \infty$ for each $n \in \mathbf{N}$, then $(\mathbf{N} \cup \{\infty\}, \leqslant)$ is an ω-cpo.
(d) $\mathbf{Pfn}(\mathbf{N}^k, \mathbf{N}^k)$ is an ω-cpo under the partial ordering $\leqslant$ of function extension. The minimal element $\perp$ is the empty function; while the lub of an ω-chain (ζ_n) of partial functions is given by

$$\Big(\bigvee_{n \geqslant 0} \zeta_n\Big)(\mathbf{a}) = \begin{cases} \zeta_n(\mathbf{a}), & \text{if } \mathbf{a} \in \mathrm{DOM}(\zeta_n) \text{ for some } n \geqslant 0; \\ \text{undefined}, & \text{if no such } n \text{ exists.} \end{cases}$$

The definition of an ω-chain ensures that if $\mathbf{a}$ is in the domain of definition of both ζ_m and ζ_n then $\zeta_m(\mathbf{a}) = \zeta_n(\mathbf{a})$, so that $\bigvee_{n \geqslant 0} \zeta_n$ is well defined.

11 Definition (Order Semantics of k-Variable **while**-Programs). We assign to every k-variable **while**-program P an interpretation $\gamma_P : \mathbf{N}^k \to \mathbf{N}^k$, obtained inductively by:

(1) Identical to (1) of Definition 1.
(2) Identical to (2) of Definition 1.

[In both (1) and (2) we substitute γ for α throughout.]

[1] Recall [e.g., from Arbib, Kfoury, and Moll (1981), Section 5.1] that a poset $(A, \leqslant)$ is a set A together with a partial order $\leqslant$, i.e., a relation on A which is *reflexive* ($a \leqslant a$ for each a in A), *antisymmetric* ($a \leqslant a'$ and $a' \leqslant a$ implies $a = a'$) and *transitive* ($a \leqslant a'$ and $a' \leqslant a''$ implies $a \leqslant a''$).

(3) If $\hat{\mathbb{S}}$ is the **while** statement

$$\textbf{while } Xu \neq Xv \textbf{ do } \mathbb{S}$$

then $\gamma_{\hat{\mathbb{S}}} = \bigvee_{n \geqslant 0} \zeta_n$ where $\zeta_0 = \perp$ and

$$\zeta_{n+1}(\mathbf{a}) = \begin{cases} \zeta_n \circ \gamma_{\mathbb{S}}(\mathbf{a}), & \text{if } [\mathbf{a}]_u \neq [\mathbf{a}]_v; \\ \mathbf{a}, & \text{if } [\mathbf{a}]_u = [\mathbf{a}]_v. \end{cases}$$

Here $[\mathbf{x}]_u$ is the uth component of k-dimensional vector $\mathbf{x}$, and $\gamma_{\mathbb{S}}$ is the (order) semantics of $\mathbb{S}$.

The lub in (3) is well defined because for all $n \geqslant 0$, $\zeta_n \leqslant \zeta_{n+1}$ as shown next, so that the sequence (ζ_n) is also an ω-chain.

12 Lemma. *In part* (3) *of Definition* 11, $\zeta_n \leqslant \zeta_{n+1}$ *for all* $n \geqslant 0$.

PROOF. By induction on n. Since $\zeta_0 = \perp$, $\zeta_0 \leqslant \zeta_1$. Assume next that $\zeta_n \leqslant \zeta_{n+1}$; we must show that $\zeta_{n+1} \leqslant \zeta_{n+2}$. Let $\mathbf{a} \in \mathbf{N}^k$, and consider the values of $\zeta_{n+1}(\mathbf{a})$ and $\zeta_{n+2}(\mathbf{a})$. If $[\mathbf{a}]_u = [\mathbf{a}]_v$, then $\zeta_{n+1}(\mathbf{a}) = \mathbf{a} = \zeta_{n+2}(\mathbf{a})$.

If $[\mathbf{a}]_u \neq [\mathbf{a}]_v$, then $\zeta_{n+1}(\mathbf{a}) = \zeta_n \circ \gamma_{\mathbb{S}}(\mathbf{a})$ and $\zeta_{n+2}(\mathbf{a}) = \zeta_{n+1} \circ \gamma_{\mathbb{S}}(\mathbf{a})$, by definition. By the induction hypothesis, $\zeta_n \leqslant \zeta_{n+1}$, so that if $\zeta_{n+1}(\mathbf{a})$ is defined, so is $\zeta_{n+2}(\mathbf{a})$—and both have the same value. □

Our next task is to relate α_P, β_P, and γ_P for any **while**-program P.

Consider the interpretation $\beta^\dagger$ of the iterate $K^\dagger$ (see Lemma 5). β_1 and β_2 are the disjoint functions induced by β. We define the sequence $(\xi_n \mid n \geqslant 0)$ by setting

$$\xi_n = \sum_{0 \leqslant j \leqslant n} \beta_2 \circ \beta_1^j,$$

which we may view as an "approximation" of the iterate $\beta^\dagger = \Sigma_{j \geqslant 0} \beta_2 \circ \beta_1^j$. Clearly $\xi_n \leqslant \xi_{n+1}$, so that (ξ_n) is an ω-chain. Note that

$$\begin{aligned} \xi_{n+1} &= \beta_2 + \beta_2 \circ \beta_1 + \cdots + \beta_2 \circ \beta_1^n + \beta_2 \circ \beta_1^{n+1} \\ &= \beta_2 + (\beta_2 + \cdots + \beta_2 \circ \beta_1^{n+1} + \beta_2 \circ \beta_1^n) \circ \beta_1 \\ &= \beta_2 + \xi_n \circ \beta_1. \end{aligned}$$

13 Lemma. *Let K be the one-entry, two-exit flow diagram of Lemma 5 with interpretation $\beta: \mathbf{N}^k \to \mathbf{N}^k + \mathbf{N}^k$. Then the semantics $\beta^\dagger$ of the iterate $K^\dagger$ satisfies*

$$\beta^\dagger = \bigvee_{n \geqslant 0} \xi_n$$

where $\xi_0 = \perp$ *and* $\xi_{n+1} = \beta_2 + \xi_n \circ \beta_1$.

PROOF. For every n, $\xi_n = \Sigma_{0 \leqslant j \leqslant n} \beta_2 \circ \beta_1^j$, so that $\xi_n \leqslant \Sigma_{j \geqslant 0} \beta_2 \circ \beta_1^j = \beta^\dagger$. Hence $\bigvee_{n \geqslant 0} \xi_n \leqslant \beta^\dagger$.

It remains to prove the opposite containment, i.e., $\beta^\dagger = \Sigma_{j\geqslant 0}\,\beta_2 \circ \beta_1^j \leqslant \bigvee_{n\geqslant 0}\xi_n$. Take an arbitrary $(\mathbf{a},\mathbf{b}) \in \beta^\dagger$. Then $(\mathbf{a},\mathbf{b}) \in \beta_2 \circ \beta_1^n$ for some $n \geqslant 0$. Hence $(\mathbf{a},\mathbf{b}) \in \xi_n$, and therefore $(\mathbf{a},\mathbf{b}) \in \bigvee_{n\geqslant 0}\xi_n$. □

14 Theorem. *For every* **while**-*program* P, $\alpha_P = \beta_P = \gamma_P$; *that is, the basic semantics, partially additive semantics, and order semantics of* **while**-*programs are identical.*

PROOF. By Theorem 7, it suffices to show $\beta_P = \gamma_P$. We prove by induction on the structure of **while**-programs that $\beta_{\mathbb{S}} = \gamma_{\mathbb{S}}$ for every statement $\mathbb{S}$.

If $\mathbb{S}$ is an assignment or a compound statement then trivially $\beta_{\mathbb{S}} = \gamma_{\mathbb{S}}$, since (1) and (2) of Definitions 6 and 11 are identical.

If $\hat{\mathbb{S}}$ is the **while** statement **while** $Xu \neq Xv$ **do** $\mathbb{S}$, we assume that $\beta_{\mathbb{S}} = \gamma_{\mathbb{S}}$ and prove that $\beta_{\hat{\mathbb{S}}} = \gamma_{\hat{\mathbb{S}}}$. By Definition 6, $\beta_{\hat{\mathbb{S}}} = \Sigma_{n\geqslant 0}\, p_F \circ (\beta_{\mathbb{S}} \circ p_T)^n$. Hence by Lemma 13,

$$\beta_{\hat{\mathbb{S}}} = \bigvee_{n\geqslant 0} \xi_n,$$

where $\xi_0 = \perp$ and $\xi_{n+1} = p_F + \xi_n \circ (\beta_{\mathbb{S}} \circ p_T)$. By the induction hypothesis, $\xi_{n+1} = p_F + \xi_n \circ (\gamma_{\mathbb{S}} \circ p_T)$. But this means

$$\xi_{n+1}(\mathbf{a}) = \begin{cases} \xi_n \circ \gamma_{\mathbb{S}}(\mathbf{a}), & \text{if } [\mathbf{a}]_u \neq [\mathbf{a}]_v; \\ \mathbf{a}, & \text{if } [\mathbf{a}]_u = [\mathbf{a}]_v. \end{cases}$$

Then by Definition 11 we have that $\xi_n = \zeta_n$ for all n. Thus $\beta_{\hat{\mathbb{S}}} = \bigvee_{n\geqslant 0}\zeta_n = \gamma_{\hat{\mathbb{S}}}$. □

We conclude this section with the "least-fixed-point" characterization of order semantics. For the next theorem we need the following notions.

15 Definition

(1) Let $(A, \leqslant)$ be a poset, and let $\Theta: A \to A$ be a total map. We say that a_0 in A is a *fixed point* of Θ if $\Theta(a_0) = a_0$, and a *least fixed point* of Θ if $a_0 \leqslant a$ for every fixed point a of Θ.
(2) Θ is said to be *monotonic* if for all $a, b \in A$, $a \leqslant b$ implies $\Theta(a) \leqslant \Theta(b)$.
(3) Suppose in addition that $(A, \leqslant)$ is an ω-cpo, and Θ is monotonic. Then for each ω-chain (a_n) in A, $(\Theta(a_n))$ is also an ω-chain. Θ is said to be *continuous* if $\Theta(\bigvee_{n\geqslant 0} a_n) = \bigvee_{n\geqslant 0}\Theta(a_n)$ for every ω-chain (a_n).

16 Theorem (The Least-Fixed-Point Theorem). *Let* $(A, \leqslant)$ *be an* ω-*cpo, and* $\Theta: A \to A$ *a continuous total map. Then* Θ *has a least fixed point, denoted* $Y(\Theta)$, *satisfying the equation* $Y(\Theta) = \bigvee_{n\geqslant 0}\Theta^n(\perp)$.

PROOF. We establish the theorem in three stages:

(a) We prove that the sequence $(\Theta^n(\perp) \mid n \geqslant 0)$ is an ω-chain, so that $Y(\Theta)$ is well defined: Since $\Theta^0(\perp) = \perp$, certainly $\Theta^0(\perp) \leqslant \Theta^1(\perp)$. But then, since Θ is monotonic, it is immediate that $\Theta^n(\perp) \leqslant \Theta^{n+1}(\perp)$ for all n, by induction on n.

(b) We prove that $Y(\Theta)$ is a fixpoint of Θ:

$$\Theta\Big(\bigvee_{n\geqslant 0} \Theta^n(\perp)\Big) = \bigvee_{n\geqslant 0} \Theta \circ \Theta^n(\perp) \text{ (since } \Theta \text{ is continuous)}$$
$$= \bigvee_{n\geqslant 1} \Theta^n(\perp)$$
$$= Y(\Theta) \text{ [since removal of initial elements from an } \omega\text{-chain does not change its lub (Exercise 6)].}$$

(c) Finally, we prove that if a is a fixpoint of Θ, then $Y(\Theta) \leqslant a$. Certainly $\perp \leqslant a$. But then by monotonicity of Θ,

$$\Theta^n(\perp) \leqslant \Theta^n(a) \text{ for each } n \geqslant 0.$$

But a is a fixpoint of Θ, and so $\Theta^n(a) = a$ for each $n \geqslant 0$. Hence $\Theta^n(\perp) \leqslant a$, and a is an upper bound for the ω-chain $(\Theta^n(\perp))$. Thus $Y(\Theta) = \bigvee_{n\geqslant 0}\Theta^n(\perp) \leqslant a$. □

We call a total map of functions to functions a *functional*. In particular, a total map from $\mathbf{Pfn}(\mathbf{N}^k, \mathbf{N}^k)$ to $\mathbf{Pfn}(\mathbf{N}^k, \mathbf{N}^k)$ is a functional.

17 Definition. We associate with every k-variable **while**-program P a functional $\Theta_P : \mathbf{Pfn}(\mathbf{N}^k, \mathbf{N}^k) \to \mathbf{Pfn}(\mathbf{N}^k, \mathbf{N}^k)$ obtained inductively by:

(1) If $\mathbb{S}$ is the assignment statement $Xu := g(Xv)$ then $\Theta_{\mathbb{S}}$ is defined by

$$\Theta_{\mathbb{S}}(\zeta) = \gamma_{\mathbb{S}}$$

for all $\zeta \in \mathbf{Pfn}(\mathbf{N}^k, \mathbf{N}^k)$, where $\gamma_{\mathbb{S}} : \mathbf{N}^k \to \mathbf{N}^k$ maps any k-tuple $(a_1, \ldots, a_k)$ to $(a_1, \ldots, a_{u-1}, g(a_v), a_{u+1}, \ldots, a_k)$.

(2) If $\mathbb{S}$ is the compound statement **begin** $\mathbb{S}_1; \ldots; \mathbb{S}_m$ **end** then $\Theta_{\mathbb{S}}$ is defined by

$$\Theta_{\mathbb{S}}(\zeta) = Y(\Theta_{\mathbb{S}_m}) \circ \cdots \circ Y(\Theta_{\mathbb{S}_2}) \circ Y(\Theta_{\mathbb{S}_1})$$

for all $\zeta \in \mathbf{Pfn}(\mathbf{N}^k, \mathbf{N}^k)$.

(3) If $\hat{\mathbb{S}}$ is the **while** statement **while** $Xu \neq Xv$ **do** $\mathbb{S}$ then $\Theta_{\hat{\mathbb{S}}}$ is defined by

$$\Theta_{\hat{\mathbb{S}}}(\zeta) = \zeta \circ Y(\Theta_{\mathbb{S}}) \circ p_T + p_F$$

where for all $(a_1, \ldots, a_k) \in \mathbf{N}^k$:

$$p_T(a_1, \ldots, a_k) = \begin{cases} (a_1, \ldots, a_k), & \text{if } a_u \neq a_v\,; \\ \perp, & \text{if } a_u = a_v\,; \end{cases}$$

$$p_F(a_1, \ldots, a_k) = \begin{cases} (a_1, \ldots, a_k), & \text{if } a_u = a_v\,; \\ \perp, & \text{if } a_u \neq a_v\,. \end{cases}$$

In the above definition we have used the least-fixed-point operator Y, which presumes that functionals associated with **while**-programs are continuous. The next lemma justifies this use of Y in the definition.

18 Lemma. *Let Θ be a total map from* $\mathbf{Pfn}(\mathbf{N}^k, \mathbf{N}^k)$ *to* $\mathbf{Pfn}(\mathbf{N}^k, \mathbf{N}^k)$.

(1) *If Θ is a "constant" functional, i.e., Θ maps all ζ in* $\mathbf{Pfn}(\mathbf{N}^k, \mathbf{N}^k)$ *to the same function γ in* $\mathbf{Pfn}(\mathbf{N}^k, \mathbf{N}^k)$, *then Θ is continuous.*
(2) *If Θ is defined by*

$$\Theta(\zeta) = \zeta \circ \theta_1 + \theta_2$$

where θ_1 and θ_2 are disjoint elements in $\mathbf{Pfn}(\mathbf{N}^k, \mathbf{N}^k)$, *then Θ is continuous.*

PROOF. Exercise 8. □

19 Corollary. *Let P be a k-variable* **while**-*program. Then the associated functional Θ_P is continuous, and its least fixed point $Y(\Theta_P)$ is precisely $\gamma_P : \mathbf{N}^k \to \mathbf{N}^k$, the semantics of P.*

PROOF. That Θ_P is continuous follows from Lemma 18. We next prove that $Y(\Theta_P) = \gamma_P$ by induction on the structure of **while**-programs.

If $\mathbb{S}$ is an assignment statement, then $\Theta_{\mathbb{S}}(\zeta) = \gamma_{\mathbb{S}}$ for all ζ in $\mathbf{Pfn}(\mathbf{N}^k, \mathbf{N}^k)$. Hence, the least (and only) fixed point of $\Theta_{\mathbb{S}}$ is $\gamma_{\mathbb{S}}$.

If $\mathbb{S}$ is the compound statement **begin** $\mathbb{S}_1; \ldots; \mathbb{S}_m$ **end** with $Y(\Theta_{\mathbb{S}_1}) = \gamma_{\mathbb{S}_1}, \ldots, Y(\Theta_{\mathbb{S}_m}) = \gamma_{\mathbb{S}_m}$ (induction hypothesis), then the least (and only) fixed point of $\Theta_{\mathbb{S}}$ is $\gamma_{\mathbb{S}_m} \circ \cdots \circ \gamma_{\mathbb{S}_2} \circ \gamma_{\mathbb{S}_1}$.

Finally, let $\hat{\mathbb{S}}$ be the **while** statement **while** $Xu \neq Xv$ **do** $\mathbb{S}$ with $Y(\Theta_{\mathbb{S}}) = \gamma_{\mathbb{S}}$. We must show that if

$$\Theta_{\hat{\mathbb{S}}}(\zeta) = \zeta \circ \gamma_{\mathbb{S}} \circ p_T + p_F$$

then $Y(\Theta_{\hat{\mathbb{S}}}) = \gamma_{\hat{\mathbb{S}}}$. By Theorem 16, $Y(\Theta_{\hat{\mathbb{S}}}) = \bigvee_{n \geq 0} \Theta_{\hat{\mathbb{S}}}^n(\perp)$. It is easily verified that for all $n \geq 0$, $\Theta_{\hat{\mathbb{S}}}^n = \zeta_n$ [where ζ_n is given in (3) of Definition 11]. Since $\gamma_{\hat{\mathbb{S}}} = \bigvee_{n \geq 0} \zeta_n$ by definition, the result follows. □

The preceding corollary implies, in particular, that every computable function is the least fixed point of a continuous functional.

Further applications of order semantics and the least-fixed-point approach will be given in the next section.

EXERCISES FOR SECTION 5.1

1. Prove that the basic semantics of **while**-programs as given in Definition 1 are equivalent to their operational semantics obtained by tracing paths through flow diagrams, as in Section 2.3. Use induction on the structure of **while**-programs.

2. Prove Proposition 3.

3. Supply the details for the proof of Theorem 7.

4. Extend each of Definition 1, Definition 6, and Definition 11, to:

   (a) **if-then** and **if-then-else** statements,
   (b) **repeat-until** statements.

   Extend the proofs of each of Theorem 7 and Theorem 14 to include cases (a) and (b).

5. Using Example 8 as a guide, prove that in each of (a), (b), and (c) below, the two statements are equivalent (i.e., have the same semantics):

   (a) **while** $\mathcal{C}$ **do** $\mathcal{S}$ and **if** $\mathcal{C}$ **then repeat** $\mathcal{S}$ **until** $\neg\mathcal{C}$
   (b) **begin while** $\mathcal{C}$ **do** $\mathcal{S}_1$; $\mathcal{S}_2$ **end** and **begin while** $\mathcal{C}$ **do** $\mathcal{S}_1$; **if** $\mathcal{C}$ **then** $\mathcal{S}_1$ **else** $\mathcal{S}_2$ **end**
   (c) **while** $\mathcal{C}$ **do** $\mathcal{S}_1$ and **begin while** $\mathcal{C}$ **do** $\mathcal{S}_1$; **while** $\mathcal{C}$ **do** $\mathcal{S}_2$ **end**

6. Let $a_0 \leqslant a_1 \leqslant \cdots \leqslant a_n \leqslant \cdots$ be an ω-chain in the ω-cpo $(A, \leqslant)$. Prove that for every $k \geqslant 0$,

$$\bigvee_{n \geqslant 0} a_n = \bigvee_{n \geqslant k} a_n .$$

7. Show that the condition for continuity in Definition 15 can be relaxed to $\Theta(\bigvee_{n \geqslant 0} a_n) \leqslant \bigvee_{n \geqslant 0} \Theta(a_n)$. That is, prove that

$$\Theta\Big(\bigvee_{n \geqslant 0} a_n\Big) \leqslant \bigvee_{n \geqslant 0} \Theta(a_n) \Leftrightarrow \Theta\Big(\bigvee_{n \geqslant 0} a_n\Big) = \bigvee_{n \geqslant 0} \Theta(a_n).$$

8. Prove Lemma 18.

9. Consider the ω-cpo $\mathbf{Pfn}(\mathbf{N}^k, \mathbf{N}^k)$ with the partial ordering $\leqslant$ of function extension.

   (a) Exhibit a continuous functional whose least fixed point is not a computable function from $\mathbf{N}^k$ to $\mathbf{N}^k$.
   (b) Show that there are uncountably many continuous functionals over $\mathbf{Pfn}(\mathbf{N}^k, \mathbf{N}^k)$.

5.2 Recursive Programs

In the language of **while**-programs, as specified in Chapter 2, recursively written programs are not permitted. However, just as PASCAL allows recursive definitions of procedures, so might we temporarily extend the language of **while**-programs to allow recursively written programs. Our first objective in this section is to prove that functions computed by **while**-programs with recursion are already computed by **while**-programs without recursion (Theorem 2 of this section). We establish this result in the context of operational semantics. Later in the section we use techniques of denotational semantics to establish further results about the translation from programs with recursion to programs without recursion.

To motivate our later presentation, consider the following recursively written program FACT, which we use to compute a unary function. (What this function turns out to be is of no concern to us at this point.)

```
procedure FACT
begin if X1 = 0 then X1 := 1 else
    begin X2 := X1;
        X1 := pred(X1);
        FACT;
        X1 := X2 * X1
    end
end
```

Only the first line in this program, namely the heading "**procedure** FACT", is not part of the regular syntax of **while**-programs. Since we want FACT to define a unary function, we agree that every time FACT is activated, an input value is assigned to variable $X1$, which is also used as the output variable. (With this convention we do not need to mention input/output parameters in the heading of FACT.) What makes the above program recursive is that its name, FACT, *recurs* inside its body. Of course if we replace this recursive call to FACT by a call to P, which is the name of any **while**-program (without recursion), then FACT itself becomes the name of a **while**-program (without recursion).

When first encountered, recursive programs appear highly suspicious since they seem to assume that the object they are trying to define (in this example, FACT) is already defined. The way out is to regard programs written recursively as shorthand. We show below how the text of FACT is to be read in order to associate with every input value a unique computation sequence, so that the operational semantics of FACT are also well defined.

If $X1 = 0$, then the above program tells us that $X1$ is to be set to 1, and the computation halts. If φ_{FACT} is the function computed by FACT, then clearly $\varphi_{\text{FACT}}(0) = 1$.

Now if $X1 \neq 0$, we first set $X2$ to the current value of $X1$, decrease $X1$ by 1 ... and then run up against a blank wall! But the prescription is clear: To proceed further in the computation, we replace the word FACT by the body of the above procedure definition. Thus, we obtain the following expanded version of FACT—call it $\text{FACT}^{(2)}$:

```
procedure FACT(2)
begin if X1 = 0 then X1 := 1 else
    begin X2 := X1;
        X1 := pred(X1);
       ┌─────────────────────────────────────┐
       │begin if X1 = 0 then X1 := 1 else    │
       │    begin (X3) := X1;                │
       │        X1 := pred(X1);              │
       │        FACT;                        │ FACT
       │        X1 := (X3) * X1              │
       │    end                              │
       │end;                                 │
       └─────────────────────────────────────┘
        X1 := X2 * X1
    end
end
```

Note one "trick": When we replaced FACT, we also changed $X2$ to $X3$ (the circled variable) so that the assignment "$X3 := X1$" would not overwrite the value of $X2$. On the other hand, we use the same $X1$ in the inner and outer bodies, since $X1$ is the input/output variable for FACT. We clearly have $\varphi_{\text{FACT}}(1) = 1$.

What about $\varphi_{\text{FACT}}(2)$? If we follow through the computation of $\text{FACT}^{(2)}$ on input 2, we again encounter the troublesome recursive call. So we replace the word FACT in $\text{FACT}^{(2)}$ by the body of the original recursive procedure, and obtain the following expanded version of $\text{FACT}^{(2)}$ (and therefore of FACT itself)—call it $\text{FACT}^{(3)}$:

```
procedure FACT(3)
begin if X1 = 0 then X1 := 1 else
    begin X2 := X1;
        X1 := pred(X1);
       ┌──────────────────────────────────────────────┐
       │begin if X1 = 0 then X1 := 1 else             │
       │    begin (X3) := X1;                         │
       │        X1 := pred(X1);                       │
       │       ┌───────────────────────────────────┐  │
       │       │begin if X1 = 0 then X1 := 1 else  │  │
       │       │    begin (X4) := X1;              │  │
       │       │        X1 := pred(X1);            │  │
       │       │        FACT;                      │ FACT │ FACT(2)
       │       │        X1 := (X4) * X1            │  │
       │       │    end                            │  │
       │       │end;                               │  │
       │       └───────────────────────────────────┘  │
       │        X1 := (X3) * X1                       │
       │    end                                       │
       │end;                                          │
       └──────────────────────────────────────────────┘
        X1 := X2 * X1
    end
end
```

Here again, the circled variables refer to "new versions" of $X2$. It is now clear that $\varphi_{\text{FACT}}(2) = 2$.

More generally, we set $\text{FACT}^{(1)}$ to be FACT itself, and for all $n > 1$ we define the nth level expansion $\text{FACT}^{(n)}$ as

```
procedure FACT^(n)
begin if X1 = 0 then X1 := 1 else
      begin X2 := X1;
            X1 := pred(X1);
            ┌─────────────────────────────────┐
            │ body of FACT^(n-1) with each    │
            │ Xi, i ≥ 2, replaced by Xi + 1   │
            └─────────────────────────────────┘
            X1 := X2 * X1
      end
end
```

It is easily verified that for all $n > a$, the computation of $\text{FACT}^{(n)}$ on input a does not reach the recursive call to FACT. Hence, for all input a and all $n > a$, the computation of $\text{FACT}^{(n)}$ on a proceeds according to our conventions for computations by **while**-programs (without recursion).

We now associate a **while**-program without recursion with each of the recursive procedures $\text{FACT}^{(n)}$, $n \geqslant 1$. Specifically, let e be an index for the empty function, obtained by setting

```
P_e = begin X1 := 1;
            while X1 ≠ 0 do begin end
      end
```

We then define the program-rewriting function $f: \mathbf{N} \to \mathbf{N}$ by

```
P_f(x) = begin if X1 = 0 then X1 := 1 else
               begin X2 := X1;
                     X1 := pred(X1);
                     ┌──────────────────────────┐
                     │ P_x with each Xi, i ≥ 2, │
                     │ replaced by Xi + 1       │
                     └──────────────────────────┘
                     X1 := X2 * X1
               end
         end
```

Hence **while**-program $\mathbf{P}_{f^n(e)}$ is identical to the body of the recursive procedure $\text{FACT}^{(n)}$, except that the recursive call to FACT is replaced by $\mathbf{P}_e$.

We conclude that for all $a \in \mathbf{N}$ and $n \geqslant 1$:

$$\varphi_{f^n(e)}(a) = \begin{cases} \varphi_{\text{FACT}^{(n)}}(a), & \text{if the computation of } \text{FACT}^{(n)} \\ & \text{on input } a \text{ does not call FACT;} \\ \perp, & \text{otherwise.} \end{cases}$$

The computation of $\mathbf{P}_{f^n(e)}$ on an arbitrary input a is thus identical to the computation of $\text{FACT}^{(n)}$ on a, as long as $\text{FACT}^{(n)}$ does not call FACT. Note also that if FACT halts on input a, then there is an $n \geqslant 1$ such that

$\text{FACT}^{(n)}$ on input a halts *and* $\text{FACT}^{(n)}$ does not invoke FACT. We can therefore simulate the computation of FACT on input a with the following (informal) algorithm:

1. Set n to 1.
2. Run **while**-program $\mathbf{P}_{f^n(e)}$ on input a (which also requires that we initialize Xi to zero, for $i \geqslant 2$).
3. If the computation of $\mathbf{P}_{f^n(e)}$ reaches a call to $\mathbf{P}_e$, increment n by 1 and go to step 2.
4. If the computation of $\mathbf{P}_{f^n(e)}$ halts "normally", i.e., without reaching $\mathbf{P}_e$, return the final value stored in $X1$ as output.

We now give a general treatment of the semantics of recursive programs.

1 Definition. A *recursive program* R is of the form

procedure R
body of R

where the "body of R" is an arbitrary **while**-program with this one exception: In the inductive definition of **while**-programs, we include the name R in the basis along with assignment statements.

For simplicity we only consider the case where R is used to compute a unary function, denoted $\varphi_R : \mathbf{N} \to \mathbf{N}$, with variable $X1$ used for both input and output. Let us also assume that the body of R contains at most k variables, $X1, X2, \ldots, Xk$. We represent R schematically by

procedure R
begin $\ldots R \ldots R \ldots \quad \ldots R \ldots$ **end**

We associate with R a program-rewriting function $f_R : \mathbf{N} \to \mathbf{N}$ defined by

$$\mathbf{P}_{f_R(x)} = \textbf{begin} \ldots \mathbf{P}_{f_1(x)} \ldots \mathbf{P}_{f_2(x)} \ldots \quad \ldots \mathbf{P}_{f_m(x)} \ldots \textbf{end}$$

where $\mathbf{P}_{f_1(x)}, \mathbf{P}_{f_2(x)}, \ldots, \mathbf{P}_{f_m(x)}$ are respectively substituted for the first, second, $\ldots$, mth (and last) occurrence of R in the procedure body. If $\mathbf{P}_x$ contains variables $X1, X2, \ldots, Xl$, then $\mathbf{P}_{f_j(x)}$ is obtained from $\mathbf{P}_x$ by replacing each variable Xi, $2 \leqslant i \leqslant l$, by Xi' where $i' = i + k - 1 + (j - 1) * (l - 1)$. These changes of variables ensure that $\mathbf{P}_x$ can use $X1$ to communicate with its environment, but will not have side effects on the other variables in its environment, $X2, \ldots, Xk$.

As in the case of FACT, we can associate with recursive program R the *nth level expansion* $R^{(n)}$, for every $n \geqslant 1$. If $\mathbf{P}_e$ is the program given earlier for the empty function, then the body of $R^{(n)}$ is none other than $\mathbf{P}_{f_R^n(e)}$ where every occurrence of $\mathbf{P}_e$ is changed to R. It is now readily verified that for all $a \in \mathbf{N}$ and $n \geqslant 1$:

$$\varphi_{f_R^n(e)}(a) = \begin{cases} \varphi_{R^{(n)}}(a), & \text{if the computation of } R^{(n)} \\ & \text{on input } a \text{ does not call } R; \\ \perp, & \text{otherwise.} \end{cases}$$

We can now simulate the computation of R on input a with the following (informal) algorithm:

1. Set n to 1.
2. Run $\mathbf{P}_{f_R^n(e)}$ on input a, which requires that we initialize Xi to zero, for all $i \geqslant 2$.
3. If the computation of $\mathbf{P}_{f_R^n(e)}$ reaches a call to $\mathbf{P}_e$, increment n by 1 and go to step 2.
4. If the computation of $\mathbf{P}_{f_R^n(e)}$ halts "normally", return the final value stored in $X1$.

2 Theorem. *For every recursive program R, there is a* **while**-*program P (without recursion) such that $\varphi_R = \varphi_P$.*

PROOF. We define a **while**-program P that implements the informal algorithm described above. To this end, we introduce a new variable FLAG which will be set to 1 whenever the computation reaches $\mathbf{P}_e$. (Of course, FLAG is Xi for some $i \geqslant 1$, but we use a different name, because we do not want FLAG to be affected by variable renaming.) We also define index u by setting

$$\mathbf{P}_u = \textbf{begin } \mathrm{FLAG} := 1 \textbf{ end}$$

and replace every occurrence of $\mathbf{P}_e$ in the informal algorithm by $\mathbf{P}_u$ to obtain:

1. Set n to 1.
2. Set FLAG to 0 and run $\mathbf{P}_{f_R^n(u)}$ on input a. If and when the computation halts, go to step 3.
3. If FLAG = 1, increment n by 1 and go to step 2.
4. If FLAG = 0, return final value stored in $X1$.

We now implement this algorithm in the form of a **while**-program. For this, we first define a new program-rewriting function $g_R : \mathbf{N} \to \mathbf{N}$ by setting

$\mathbf{P}_{g_R(x)} =$ **begin** FLAG := 0;

  | $\mathbf{P}_{f_R(x)}$ where every test $\mathcal{C}$ in a **while** statement is changed to $\mathcal{C}$ FLAG = 0 |

  if FLAG = 0 **then** $X1 := succ(X1)$

      else $X1 := 0$

  end

It is easily verified that for all $a \in \mathbf{N}$ and $n \geqslant 1$:

$$\varphi_{g_R^n(u)}(a) = \begin{cases} \varphi_{f_R^n(e)}(a) + 1, & \text{if the computation of } \mathbf{P}_{f_R^n(e)} \text{ on input } a \text{ does not reach a call to } \mathbf{P}_e; \\ 0, & \text{otherwise.} \end{cases}$$

As in Section 3.2, $\Phi : \mathbf{N}^2 \to \mathbf{N}$ is the universal function for the unary computable functions. The desired **while**-program P can now be put in the following form (using macros):

$$\begin{aligned}&\textbf{begin } n := 0;\\&\qquad \textbf{repeat } n := n + 1 \textbf{ until } \Phi(g_R^n(u), X1) \neq 0;\\&\qquad X1 := \Phi(g_R^n(u), X1) \dot{-} 1\\&\textbf{end}\end{aligned}$$

It is clear that $\varphi_P = \varphi_R$. □

THE FIRST RECURSION THEOREM

So far in this section we have used recursive programs to compute functions from $\mathbf{N}$ to $\mathbf{N}$. We can also use recursion to compute functions from $\mathbf{N}^j$ to $\mathbf{N}$, for some $j \geqslant 1$, as well as functions from $\mathbf{N}^k$ to $\mathbf{N}^k$ (where k is the number of variables appearing in the program). In each case, the definition of the program-rewriting function f_R (induced by recursive program R) and the proof of Theorem 2 must be slightly altered (Exercise 5 of this section).

The tools of denotational semantics are especially convenient when we view programs as "vector processors", which transform j-dimensional vectors to j-dimensional vectors, for some fixed $j \geqslant 1$. For simplicity, and with no loss of generality, we restrict our attention to $j = 1$.

Then, let R be a recursive program as in Definition 1, which is used to compute a unary function:

$$\begin{aligned}&\textbf{procedure } R\\&\textbf{begin} \ldots R \ldots R \ldots \quad \ldots R \ldots \textbf{end}\end{aligned}$$

We wish to associate a continuous functional $\Theta_R : \mathbf{Pfn}(\mathbf{N},\mathbf{N}) \to \mathbf{Pfn}(\mathbf{N},\mathbf{N})$ with recursive program R, which will characterize the semantics of R; that is, Θ_R will satisfy the relation $Y(\Theta_R) = \varphi_R$. Such a Θ_R cannot be defined inductively, at least not directly, given our definition of the functional Θ_P associated with a **while**-program P (Definition 17 of Section 5.1).[2]

To avoid undue complications, we define the functional Θ_R associated with a recursive program R directly, without appealing to Definition 17 of Section 5.1. Given an arbitrary function ζ in $\mathbf{Pfn}(\mathbf{N},\mathbf{N})$, we define the following **while**-program using ζ as a macro

$$\textbf{begin} \ldots X1 := \zeta(X1) \ldots X1 := \zeta(X1) \ldots \quad \ldots X1 := \zeta(X1) \ldots \textbf{end},$$

[2] In general, an inductive definition of Θ_R requires setting up a system of simultaneous functional equations, say $\zeta_1 = \Theta_1(\zeta_1, \ldots, \zeta_n), \ldots, \zeta_n = \Theta_n(\zeta_1, \ldots, \zeta_n)$, which is the result of replacing every iterative statement by a conditional statement that calls itself recursively. The least fixed point of the system $Y(\Theta_1, \Theta_2, \ldots, \Theta_n)$ gives us the simultaneous denotations of R and the recursive programs in terms of which R is now written. Details of this approach can be found in Section 5.3 of deBakker (1980).

which is obtained by replacing every call to R in the body of R by "$X1 := \zeta(X1)$". The resulting **while**-program computes a function $\xi \in \mathbf{Pfn}(\mathbf{N}, \mathbf{N})$ which we take to be the value of Θ_R at ζ, i.e., $\Theta_R(\zeta) = \xi$.

3 Lemma. *Let R be a recursive program, and let Θ_R be the associated functional on* $\mathbf{Pfn}(\mathbf{N}, \mathbf{N})$. *Then Θ_R is continuous.*

PROOF. We first observe that Θ_R is monotonic. That is, for all ζ, $\zeta' \in \mathbf{Pfn}(\mathbf{N}, \mathbf{N})$,

$$\zeta \leqslant \zeta' \Rightarrow \Theta_R(\zeta) \leqslant \Theta_R(\zeta').$$

Hence if the sequence of functions $\zeta_0, \zeta_1, \ldots, \zeta_n, \ldots$ is an ω-chain, so is the sequence $\Theta_R(\zeta_0), \Theta_R(\zeta_1), \ldots, \Theta_R(\zeta_n), \ldots$. For all $i \geqslant 0$ we have

$$\zeta_i \leqslant \bigvee_{n \geqslant 0} \zeta_n$$

so that by monotonicity of Θ_R, we also have

$$\Theta_R(\zeta_i) \leqslant \Theta_R\Big(\bigvee_{n \geqslant 0} \zeta_n\Big)$$

Hence $\bigvee_{n \geqslant 0} \Theta_R(\zeta_n) \leqslant \Theta_R(\bigvee_{n \geqslant 0} \zeta_n)$.

We next prove the opposite inclusion, $\Theta_R(\bigvee_{n \geqslant 0} \zeta_n) \leqslant \bigvee_{n \geqslant 0} \Theta_R(\zeta_n)$. That is, for all ordered pairs (a, b), we show that $(a, b) \in \Theta_R(\bigvee_{n \geqslant 0} \zeta_n)$ implies $(a, b) \in \bigvee_{n \geqslant 0} \Theta_R(\zeta_n)$. (We may think of a unary function as a set of ordered pairs.) Consider the text of R:

procedure R
begin $\ldots R \ldots R \ldots \quad \ldots R \ldots$ **end**

where R occurs m times in the procedure body, for some $m \geqslant 1$. A **while**-program that computes the function $\Theta_R(\bigvee_{n \geqslant 0} \zeta_n)$ is

$$\textbf{begin} \ldots X1 := \bigvee_{n \geqslant 0} \zeta_n(X1) \ldots X1 := \bigvee_{n \geqslant 0} \zeta_n(X1) \ldots$$
$$\ldots X1 := \bigvee_{n \geqslant 0} \zeta_n(X1) \ldots \textbf{end}$$

We assume that this program converges on input a, and returns output b. This means that in the computation sequence corresponding to input a, every instance of the instruction "$X1 := \bigvee_{n \geqslant 0} \zeta_n(X1)$" can be replaced by "$X1 := \zeta_i(X1)$" for some $i \geqslant 0$, without changing the outcome of the computation. Hence, making this substitution for each of the m occurrences of "$X1 := \bigvee_{n \geqslant 0} \zeta_n(X1)$", we obtain a program

$$\textbf{begin} \ldots X1 := \zeta_{i_1}(X1) \ldots X1 := \zeta_{i_2}(X1) \ldots \quad \ldots X1 := \zeta_{i_m}(X1) \ldots \textbf{end}$$

which converges on input a, and returns output b. The same conclusion therefore holds for the following program

$$\textbf{begin} \ldots X1 := \zeta_j(X1) \ldots X1 := \zeta_j(X1) \ldots \quad \ldots X1 := \zeta_j(X1) \ldots \textbf{end}$$

where $j = max(i_1, \ldots, i_m)$. But this last program computes the function $\Theta_R(\zeta_j)$, so that $(a,b) \in \Theta_R(\zeta_j)$. This implies that $(a,b) \in \bigvee_{n\geqslant 0}\Theta_R(\zeta_n)$, as desired.

We conclude that $\bigvee_{n\geqslant 0}\Theta_R(\zeta_n) = \Theta_R(\bigvee_{n\geqslant 0}\zeta_n)$, and Θ_R is therefore continuous. □

By Theorem 16 of Section 5.1 we know that every continuous functional Θ has a least fixed point $Y(\Theta)$. What then is the relationship between φ_R and $Y(\Theta_R)$, i.e., between the semantics of recursive program R and the least fixed point of its associated functional Θ_R? We now prove that they are one and the same.

4 Lemma. *If R is a recursive program and Θ_R is the associated continuous functional, then $\varphi_R = Y(\Theta_R)$.*

Proof. Let $f_R : \mathbf{N} \to \mathbf{N}$ be the program-rewriting function induced by R, which we introduced after Definition 1. We first prove that for all computable functions φ_i (the function computed by **while**-programs $\mathbf{P}_i$) we have: $\varphi_{f_R(i)} = \Theta_R(\varphi_i)$.

We represent recursive program R schematically by.

procedure R
begin . . . R . . . R . . .   . . . R . . . **end**

$\mathbf{P}_{f_R(i)}$ was obtained from $\mathbf{P}_i$ by substituting $\mathbf{P}_{f_j(i)}$ for the jth occurrence of R in the body of R:

$$\mathbf{P}_{f_R(i)} = \textbf{begin} \ldots \mathbf{P}_{f_1(i)} \ldots \mathbf{P}_{f_2(i)} \ldots \quad \ldots \mathbf{P}_{f_m(i)} \ldots \textbf{end}$$

which computes the same function as **while**-program

$$\textbf{begin} \ldots X1 := \Phi(i, X1) \ldots X1 := \Phi(i, X1) \ldots$$
$$\ldots X1 := \Phi(i, X1) \ldots \textbf{end}$$

By the definition of Θ_R, this program computes $\Theta_R(\varphi_i)$.

The preceding argument is easily adapted to show by induction on n that $\varphi_{f_R^n(i)} = \Theta_R^n(\varphi_i)$, for all i and n. [We take $f_R^0(i) = i$ and $\Theta_R^0(\varphi_i) = \varphi_i$.] In particular, $\varphi_{f_R^n(e)} = \Theta_R^n(\perp)$ for all n, and where e is an index for the empty function $\perp \in \mathbf{Pfn}(\mathbf{N},\mathbf{N})$.

By Theorem 16 of Section 5.1, $\bigvee_{n\geqslant 0}\Theta_R^n(\perp)$ is precisely the least fixed point $Y(\Theta_R)$. By the discussion preceding Theorem 2, $\varphi_{f_R^0(e)}, \varphi_{f_R^1(e)}, \ldots, \varphi_{f_R^n(e)}, \ldots$ is an ω-chain and φ_R is precisely $\bigvee_{n\geqslant 0}\varphi_{f_R^n(e)}$. □

The First Recursion Theorem (also called the Weak Recursion Theorem) due to Kleene is a restatement of Theorem 16 of Section 5.1 in the context of the computable functions.

5 Theorem (The First Recursion Theorem[3]). *Let $f_R : \mathbf{N} \to \mathbf{N}$ be the program-rewriting (total computable) function induced by a recursive program R. Then there is an index m such that*

(1) $\varphi_m = \varphi_{f_R(m)}$ = *the function* φ_R *computed by R, and,*
(2) *for all n, if* $\varphi_n = \varphi_{f_R(n)}$, *then* $\varphi_m \leqslant \varphi_n$.

PROOF. In the proof of Lemma 4, we showed that $\varphi_{f_R(i)} = \Theta_R(\varphi_i)$ for all i. Hence if φ_m is the least fixed point of Θ_R, then φ_m must satisfy (1) and (2). □

6 Example. For the recursive program FACT considered earlier in this section, the corresponding functional $\Theta_{\text{FACT}} : \mathbf{Pfn(N, N)} \to \mathbf{Pfn(N, N)}$ is

$$\Theta_{\text{FACT}}(\zeta)(x) = \begin{cases} 1, & \text{if } x = 0; \\ x * \zeta(x-1), & \text{if } x \neq 0. \end{cases}$$

It is easily seen that the familiar factorial function $x!$ is a least fixed point of Θ_{FACT}—in fact, it is also the only solution of the equation $\zeta = \Theta_{\text{FACT}}(\zeta)$ since $x!$ is total and no function can extend it.

7 Example. Recursive program EVEN uses only one variable, $X1$, both for input and output.

```
procedure EVEN
begin if X1 = 0 then X1 := 1
    else if X1 = 1 then
        begin X1 := 3; EVEN end
        else begin X1 := X1 - 2; EVEN end
end
```

This program induces $\Theta_{\text{EVEN}} : \mathbf{Pfn(N, N)} \to \mathbf{Pfn(N, N)}$ according to the definition

$$\Theta_{\text{EVEN}}(\zeta)(x) = \begin{cases} 1, & \text{if } x = 0; \\ \zeta(3), & \text{if } x = 1; \\ \zeta(x-2), & \text{otherwise.} \end{cases}$$

We then see that the least fixed point of Θ_{EVEN}, and thus the function actually computed by EVEN, is the function $even : \mathbf{N} \to \mathbf{N}$ given by

$$even(x) = \begin{cases} 1, & \text{if } x \text{ is even;} \\ \perp, & \text{if } x \text{ is odd.} \end{cases}$$

[3] Theorem 5 is only a special case of Kleene's original theorem, which is often stated in terms of more general program-rewriting functions, the so-called *extensional* functions. Extensional functions correspond to "recursive operators", which are more general than the functionals considered here (i.e., Θ_R for some recursive program R). The interested reader is referred to Section 11.5 of Rogers (1967).

In fact, any ζ satisfying

$$\zeta(x) = \begin{cases} 1, & \text{if } x \text{ is even;} \\ \zeta(1), & \text{if } x \text{ is odd} \end{cases}$$

is a fixed point of Θ_{EVEN}.

8 Example. Define recursive program ANY as follows:

procedure ANY
begin ANY; $X1 := X1$ **end**

ANY induces the functional $\Theta_{\text{ANY}} : \mathbf{Pfn(N, N)} \to \mathbf{Pfn(N, N)}$ given by

$$\Theta_{\text{ANY}}(\zeta)(x) = \zeta(x) \qquad \text{for all } x.$$

The least fixed point of Θ_{ANY} is thus $\bigvee_{n \geqslant 0} \Theta^n_{\text{ANY}}(\perp) = \bigvee_{n \geqslant 0} \perp = \perp$, the empty function. However, any $\zeta : \mathbf{N} \to \mathbf{N}$ is also a fixed point of Θ_{ANY}.

9 Example. Corresponding to recursive program EMPTY

procedure EMPTY
begin EMPTY; $X1 := succ(X1)$ **end**

we have the functional $\Theta_{\text{EMPTY}} : \mathbf{Pfn(N, N)} \to \mathbf{Pfn(N, N)}$ with

$$\Theta_{\text{EMPTY}}(\zeta)(x) = succ(\zeta(x)).$$

Hence ζ is a fixed point of Θ_{EMPTY} just in case $\zeta(x) = succ(\zeta(x))$ for every $x \in \text{DOM}(\zeta)$. This is a contradiction, unless $\text{DOM}(\zeta)$ is empty. Hence $\perp$ is the least and only fixed point of Θ_{EMPTY}.

We conclude this section with a few remarks on a natural generalization of the results above. The reader familiar with recursive programming will have noticed that Definition 1 is rather restrictive, in that it does not allow for a system of simultaneously recursive programs. The general version of Definition 1 would define a *system of recursive programs* to be a system of recursive definitions of the form:

procedure R_1
begin $\ldots R_{11} \ldots R_{12} \ldots \quad \ldots R_{1n_1}$ **end**

procedure R_2
begin $\ldots R_{21} \ldots R_{22} \ldots \quad \ldots R_{2n_2}$ **end**

procedure R_m
begin $\ldots R_{m1} \ldots R_{m2} \ldots \quad \ldots R_{mn_m}$ **end**

where every $R_{ij} \in \{R_1, \ldots, R_m\}$. To "solve" such a sytem means to determine **while**-programs $P_1, \ldots, P_m$ (without recursion) that compute the same functions as $R_1, \ldots, R_m$, respectively. For simplicity, we can take all the programs in $\{R_1, \ldots, R_m\}$ to compute unary functions.

Corresponding to such a system of recursive programs, we have the following system of simultaneous equations:

$$\zeta_1 = \Theta_{R_1}(\zeta_1, \ldots, \zeta_m),$$
$$\zeta_2 = \Theta_{R_2}(\zeta_1, \ldots, \zeta_m),$$
$$\ldots$$
$$\zeta_m = \Theta_{R_m}(\zeta_1, \ldots, \zeta_m),$$

where $\Theta_{R_1}, \ldots, \Theta_{R_m}$ are now induced by the definitions of $R_1, \ldots, R_m$, respectively. We may again apply the theory of least fixed points, but this time to the ω-cpo $\mathbf{Pfn}(\mathbf{N},\mathbf{N}) \times \cdots \times \mathbf{Pfn}(\mathbf{N},\mathbf{N})$ (m times), where we now use the ordering

$$(\zeta_1, \ldots, \zeta_m) \leqslant (\zeta_1', \ldots, \zeta_m') \Leftrightarrow \zeta_i \leqslant \zeta_i' \quad \text{for} \quad 1 \leqslant i \leqslant m.$$

Exercises for Section 5.2

1. Analyze recursive program EVERY defined by;

```
procedure EVERY
begin X1 := succ(X1); EVERY end
```

Are the solutions obtained here the same as those obtained in the analysis of recursive program EMPTY of Example 9?

2. Give an analysis of recursive program POWER:

```
procedure POWER
begin if X1 = 0 then X1 := 1 else
    if X1 = 1 then X1 := 2 else
        begin X1 := X1 div 2;
            POWER;
            X1 := X1 + X1
        end
end
```

Variable $X1$ is used for both input and output, and the function *power* to be computed by this program is unary.

3. Analyze recursive program SQUARE:

```
procedure SQUARE
begin if X1 ≠ 0 then
    begin X2 := pred(2 * X1);
          X1 := pred(X1);
          SQUARE;
          X1 := X1 + X2
    end
end
```

The function to be computed by SQUARE is unary, with $X1$ used for both input and output.

4. (For readers familiar with the notion of a "stack".) We wish to repeat the proof of Theorem 2, by way of introducing "programs with stacks". We define a **stack**-program to be a **while**-program—except that we allow two additional forms of assignment instructions: "push instructions" and "pop instructions". A *push instruction* is of the form

$$\textbf{stack} := Xi$$

which "pushes" the value of variable Xi on top of the stack, leaving the value of Xi unchanged. A *pop instruction* is of the form

$$Xi := \textbf{stack}$$

which "pops" out the value on top of the stack and assigns it to variable Xi. If **stack** is empty, a pop instruction has no effect. Note that a **stack**-program can use at most one stack, namely, **stack**.

For simplicity, assume all programs are used to compute *unary* functions.

(a) If R is a recursive program, show the existence of a **stack**-program S which computes the same function as R, i.e., $\varphi_R = \varphi_S$.

(b) If S is a **stack**-program, use the pairing and projection functions of Section 3.4 to show the existence of a **while**-program P (without recursion and without a stack) which computes the same function as S, i.e., $\varphi_S = \varphi_R$.

5. Recursive programs can be used to compute functions from $\mathbf{N}^j$ to $\mathbf{N}$, for $j \geqslant 1$.

(1) If R is a k-variable recursive program which is used to compute a function $\varphi_R^{(j)} : \mathbf{N}^j \to \mathbf{N}$, for some fixed $j \geqslant 1$, alter the definition of the program-rewriting function $f_R : \mathbf{N} \to \mathbf{N}$ accordingly. State and prove the corresponding generalization of Theorem 2.

(2) Suppose k-variable recursive program R is now used to compute a function $\varphi_R^{(j)} : \mathbf{N}^j \to \mathbf{N}^j$, for some fixed $j \geqslant 1$; i.e., R is now used as a "vector processor". Discuss additional changes (if any) that must be introduced in the definition of $f_R : \mathbf{N} \to \mathbf{N}$ and the proof of Theorem 2.

6. Consider a system of two recursive programs, R_1 and R_2. For simplicity, assume that R_1 and R_2 are both used to compute unary functions. For this system of two simultaneous recursive programs:

(1) State and prove the counterpart of Theorem 2.
(2) State and prove the counterparts of Lemma 3 and Lemma 4.
(3) State and prove the counterpart of Theorem 5.

7. Analyze recursive program GCD

```
procedure GCD
begin if X2 ≠ 0 then
    begin X3 := X1;
        X1 := X2;
        X2 := X3 mod X2;
        GCD
    end
end
```

We use GCD to compute a function from $\mathbf{N}^2$ to $\mathbf{N}$. As usual, inputs are stored in $X1$ and $X2$, and outputs are retrieved from $X1$.

8. Give an analysis of the following simultaneously defined recursive programs which are used to compute unary functions:

```
procedure R_1
begin if X1 ⩽ 3 then X1 := 0
    else begin X1 := succ(X1); R_1 end
end

procedure R_2
begin if X1 = 0 then X1 := 1
    else begin R_1; R_2 end
end
```

What are φ_{R_1} and φ_{R_2}?

9. Give an analysis of the following simultaneously defined recursive programs:

```
procedure R_1
begin if X1 = 1 then X1 := 0
    else if X1 mod 2 = 0 then X1 := X1 div 2
        else begin X1 := succ(X1); R_1; R_2 end
end

procedure R_2
begin if X1 = 1 then X1 := 0
    else if X1 mod 2 = 0 then X1 := X1 div 2
        else begin X1 := succ(X1); R_1; R_2 end
end
```

10. Verify, in each of the cases below, that recursive program R computes function g. (*Hint*: Show that g is the least fixed point of the functional Θ_R.)

(1) $g(x) = 2^x$.

```
procedure R
begin if X1 = 0 then X1 := 1
    else begin X1 := pred(X1); R; X1 := 2 * X1 end
end
```

(2) $g(x) = \begin{cases} 91, & \text{if } x \leqslant 100; \\ x - 10, & \text{if } x > 100. \end{cases}$

```
procedure R
begin if X1 > 100 then X1 := X1 ∸ 10
    else begin X1 := X1 + 11; R; R end
end
```

(3) $g(x_1, x_2) = x_1 \textbf{ mod } x_2$.

```
procedure R
begin if X1 ⩾ X2 then
    begin X1 := X1 ∸ X2; R end
end
```

(4) $g(x_1, x_2) = x_1$ **mod** x_2.

```
procedure R
begin if X1 ≠ 0 then
      begin X1 := pred(X1);
            R;
            X3 := succ(X1);
            if X2 = X3 then X1 := 0 else X1 := X3
      end
end
```

5.3 Proof Rules for Program Properties

The general method of the two preceding sections consisted of assigning semantic functions to programs and then reasoning about the resulting interpretations (or denotations) to establish various program properties. It is also possible to prove program properties by using a logical system specifically designed for this purpose. A system of this sort leads to what is called a *logical semantics* (or *axiomatic semantics*) of programs.

In logical semantics, programs are specified by means of formulas of the form

$$\{A_1\}\mathbb{S}\{A_2\}$$

where $\mathbb{S}$ is a program statement (possibly an entire program), and A_1 and A_2 are conditions about computation states before and after execution of $\mathbb{S}$. The intended interpretation of such a formula is:

> *If A_1 is satisfied by a given input state and if in addition $\mathbb{S}$ terminates on that input state, then A_2 will be satisfied by the resulting output state.*

An example of a program specification as just described is

$$\{X \geqslant 100\} \textbf{ while } X > 10 \textbf{ do } X := pred(X)\ \{X = 10\}.$$

This is an assertion about the input-output behavior of the statement **while** $X > 10$ **do** $X :=$ pred(X). Namely, it asserts that if an input state satisfies the condition "$X \geqslant 100$", and if the **while** statement terminates on that input state, then the output state satisfies the condition "$X = 10$". Clearly this assertion is true of **while** $X > 10$ **do** $X := pred(X)$. The following assertion, on the other hand, is false:

$$\{X \geqslant 0\} \textbf{ while } X > 10 \textbf{ do } X := pred(X)\ \{X = 10\}$$

Now consider:

$$\{X \geqslant 0\} \textbf{ while } X > 10 \textbf{ do } X := pred(X)\ \{X \leqslant 10\}$$

This is a true assertion which can be shown, by the techniques to be developed in this section, to imply the first one.

Given a program specification of the form $\{A_1\}\mathbb{S}\{A_2\}$, we call A_1 its *precondition* and A_2 its *postcondition*. Pre- and postconditions, in general, can involve well-defined notation which is not part of the programming language. For example, we may write

$$\{x > 2\}$$

$$\textbf{while } x > 2 \textbf{ do } x := succ(x)$$

$$\{\exists a, b, c \in \mathbf{N} \text{ such that } a^x + b^x = c^x\}$$

as a formal program specification (which also happens to be a true assertion—why?) or we may write

$$\{x \geq 0\}\ \textbf{begin end}\ \{\mathbf{P}_x \text{ halts on input } x\}$$

(which is false—why?). For our purposes we shall only consider pre- and postconditions which can be written as *well-formed formulas* (wff) of first-order arithmetic $\mathfrak{F}$. $\mathfrak{F}$ is the set of first-order statements about **N** involving number-theoretic operations (such as $+$, $*$, *succ*, etc.), relations (such as $=$, $<$, etc.), as well as propositional connectives and the quantifiers $\forall$ and $\exists$.[4]

1 Definition. A *program specification* is an expression of the form $\{A_1\}\mathbb{S}\{A_2\}$ where A_1 and A_2 are wff's in $\mathfrak{F}$, and $\mathbb{S}$ is a statement (possibly a program) in the language of **while**-programs. A program specification is also called a *partial correctness assertion*.

The expression $\{A_1\}\mathbb{S}\{A_2\}$ means that if A_1 is true before $\mathbb{S}$ is executed, then A_2 must hold if and when $\mathbb{S}$ terminates. This tells us that $\{A_1\}\mathbb{S}\{A_2\}$ is trivially true if $\mathbb{S}$ does not terminate. We thus say that $\{A_1\}\mathbb{S}\{A_2\}$ expresses only the partial correctness of $\mathbb{S}$, where a *partially correct program* is one which is guaranteed to deliver the desired result *if it terminates*. But whether it will actually terminate for all input values satisfying A_1 is another matter. If we can show, in addition, that the program terminates for all input values satisfying the precondition A_1, then we say that the program is *totally correct*. We first concentrate on partial correctness (the program is correct if it terminates). Later we consider ways of proving that a program does terminate for data satisfying A_1. Of course, since the Halting Problem is unsolvable, these methods are ineffective for some programs. Pragmatically, if we integrate the analysis of algorithms with their development, we can usually stay out of the domain of undecidability.

[4] For a brief introduction to the syntax and semantics of first-order predicate logic, the reader is referred to Section 4.2 of Arbib, Kfoury, and Moll (1981). See also Section 7.3 in this book.

2 Example. The following program specification involves a **while**-program with four variables X, Y, V, and W:

$\{(X \geqslant 0) \wedge (Y > 0)\}$
begin $V := 0;$
  $W := X;$
  while $W \geqslant Y$ **do**
  begin $W := W \dot{-} Y;$
    $V := V + 1$
  end
end
$\{(X = V * Y + W) \wedge (0 \leqslant W < Y)\}$

This specification asserts that for every input state satisfying the precondition $(X \geqslant 0) \wedge (Y > 0)$, the resulting output state satisfies the postcondition $(X = V * Y + W) \wedge (0 \leqslant W < Y)$, if and when the program terminates. If we can prove that the above specification is true, we will thus also establish that the program performs integer division of $X \geqslant 0$ by $Y > 0$—storing the quotient in V and the remainder in W.

We must now develop a logical system of *axioms* and *inference rules* that will enable us to verify in a systematic way that a program P meets a specification $\{A_1\}P\{A_2\}$.

Axioms and proof rules

Rules of inference, also called *proof rules*, are patterns of reasoning that will allow us to derive properties of programs. Proof rules are usually written in the form

$$\frac{F_1, F_2, \ldots, F_n}{F}$$

which is a shorthand for: Given that formulas $F_1, F_2, \ldots, F_n$ are already derived, we can also derive formula F. In the context of this section, $F_1, F_2, \ldots, F_n$, and F are each either a wff of first-order arithmetic or a partial correctness assertion.

A proof rule as described above is said to be *sound* if the truth of $F_1, \ldots, F_n$ implies the truth of F; that is, sound proof rules preserve truth.

Proof rules can be used only to derive new formulas from previously derived formulas. To start the process we need *axioms*, formulas that need not be derived from other assertions. If we choose each of our axioms to be a true wff of $\mathcal{F}$ or a true partial correctness assertion, the soundness of the proof rules will guarantee that we only derive true assertions.

We take our set of axioms, AX, to be the set of all true wff's of first-order arithmetic:

$$\mathrm{AX} = \{A \mid A \text{ is a wff in } \mathcal{F} \text{ and } A \text{ is true}\}.$$

That is, when we derive program properties (written as partial correctness assertions), we shall help ourselves to all the true first-order arithmetical wff's we may need. Hence before attempting to use a wff A in deriving some program property, we must first find out whether A is true or false. In general (unfortunately), there is no effective way to decide whether an arbitrary wff A of $\mathcal{F}$ is true, by Gödel's Incompleteness Theorem (Section 7.3). In actual practice however, the wff's we introduce in a derivation are quite simple, usually quantifier-free, wff's which are easily recognized to be true or false—and the unsolvability of the set AX is therefore avoided.

Now consider the simplest form of program statement, namely the *null statement*, which has no effect upon the values of any of the program variables. For any wff A, the following is a true partial correctness assertion:

$$\{A\}\ \textbf{begin end}\ \{A\}. \tag{1}$$

Consider next an assignment statement $X := t$, which sets X to the value of the term t. We take a term t to be an expression obtained by composing 0, *succ*, *pred*, and any of the arithmetical operations described in Proposition 1 of Section 2.2. Then for any wff B, the following is a true partial correctness assertion

$$\{B[t/X]\}\ X := t\ \{B\}, \tag{2}$$

where $B[t/X]$ is the wff obtained from B by replacing each occurrence of variable X by term t. For example, consider the assignment $X := (V + W)$ **div** 2. If B is the condition $V \leqslant X \leqslant W$, the following program specification is true:

$$\{V \leqslant (V + W)\ \textbf{div}\ 2 \leqslant W\}\ X := (V + W)\ \textbf{div}\ 2\ \{V \leqslant X \leqslant W\}$$

We can represent this situation in the form of a diagram:

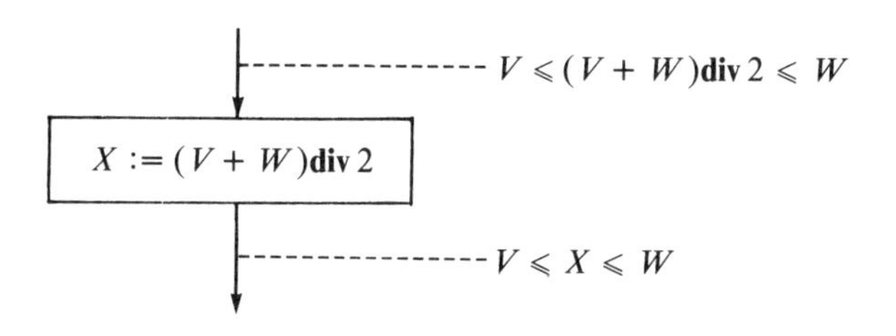

which expresses the fact that for all values assigned to V, W, and X, if the precondition $V \leqslant (V + W)$ **div** $2 \leqslant W$ is true then so is the postcondition $V \leqslant X \leqslant W$.

We now turn from axioms to rules of inference. To start with, the preceding considerations can be generalized as follows. Let A and B be wff's, considered in conjunction with the assignment statement $X := t$. If we can derive the wff $A \supset B[t/X]$, then we want to be able to derive the partial correctness assertion $\{A\}X := t\{B\}$. We state this in the form of a proof rule—the *assignment rule*:

$$\frac{A \supset B[t/X]}{\{A\}X := t\{B\}} \quad \text{(R1)}$$

The justification for (R1), i.e., for its soundness, is simple.[5] If $A \supset B[t/X]$ is true and A is true before $X := t$, then $B[t/X]$ is also true—so that B is true after $X := t$. Note that (2) follows from R1 and the axiom $A \supset A$ on taking A to be just $B[t/X]$.

Two other simple rules of inference are formulated in (R2) and (R3). The first states that if the execution of program statement $\mathbb{S}$ ensures the truth of the wff B, then it also ensures the truth of every wff implied by B.

$$\frac{\{A\}\mathbb{S}\{B\},\ B \supset C}{\{A\}\mathbb{S}\{C\}} \quad \text{(R2)}$$

The second rule states that if B is known to be a precondition for a program statement $\mathbb{S}$ to produce the result C after completion of execution of $\mathbb{S}$, then so is any other condition which implies B:

$$\frac{A \supset B,\ \{B\}\mathbb{S}\{C\}}{\{A\}\mathbb{S}\{B\}} \quad \text{(R3)}$$

The rules (R2) and (R3) are called the *consequence rules*.

Suppose we now wish to establish that $\{A\}\mathbb{S}\{B\}$ holds when $\mathbb{S}$ is a structured program statement. What we need is a rule for every type of composition of statements which allows us to infer the properties of the composite (structured) statement on the basis of the established properties of its components. To motivate the proof rule for the compound statement **begin** $\mathbb{S}_1; \mathbb{S}_2; \ldots; \mathbb{S}_n$ **end**, consider the case when $n = 2$:

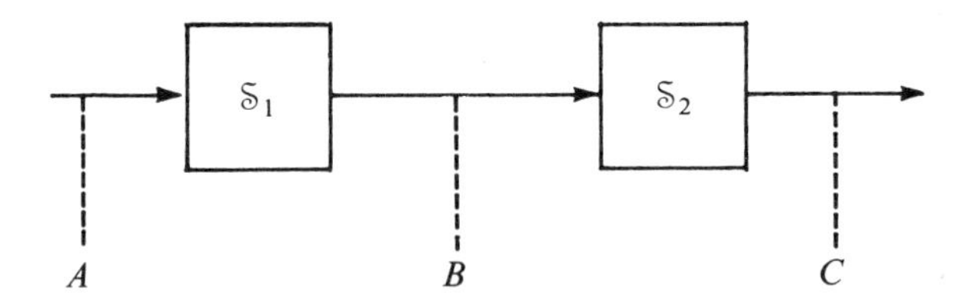

If $\{A\}\mathbb{S}_1\{B\}$ and $\{B\}\mathbb{S}_2\{C\}$ hold, then $\{A\}$ **begin** $\mathbb{S}_1; \mathbb{S}_2$ **end** $\{C\}$ holds. Formally, this rule may be expressed as follows:

$$\frac{\{A\}\mathbb{S}_1\{B\},\ \{B\}\mathbb{S}_2\{C\}}{\{A\}\ \textbf{begin}\ \mathbb{S}_1; \mathbb{S}_2\ \textbf{end}\ \{C\}} \quad \text{(R4)}$$

[5] A formal proof for the soundness of (R1) and the other inference rules can be found in Chapters 2 and 3 of deBakker (1980).

The rule (R4) generalizes in the following way, called the *compound statement rule*:

$$\frac{\{A_{i-1}\}\mathcal{S}_i\{A_i\} \text{ for } i = 1, \ldots, n}{\{A_0\} \textbf{ begin } \mathcal{S}_1 ; \mathcal{S}_2 ; \ldots ; \mathcal{S}_n \textbf{ end } \{A_n\}} \qquad \text{(R4')}$$

Note that (1) is recaptured from (R4′) by considering the case $n = 0$, on taking the null conjunction of premises to have the truth value *true*, which belongs to AX.

We next consider conditional statements. If $\mathcal{S}_1$ and $\mathcal{S}_2$ are statements and $\mathcal{C}$ is a condition (in this case a quantifier-free wff), we have a macro for the conditional statement

$$\textbf{if } \mathcal{C} \textbf{ then } \mathcal{S}_1 \textbf{ else } \mathcal{S}_2$$

which may be represented graphically by the diagram

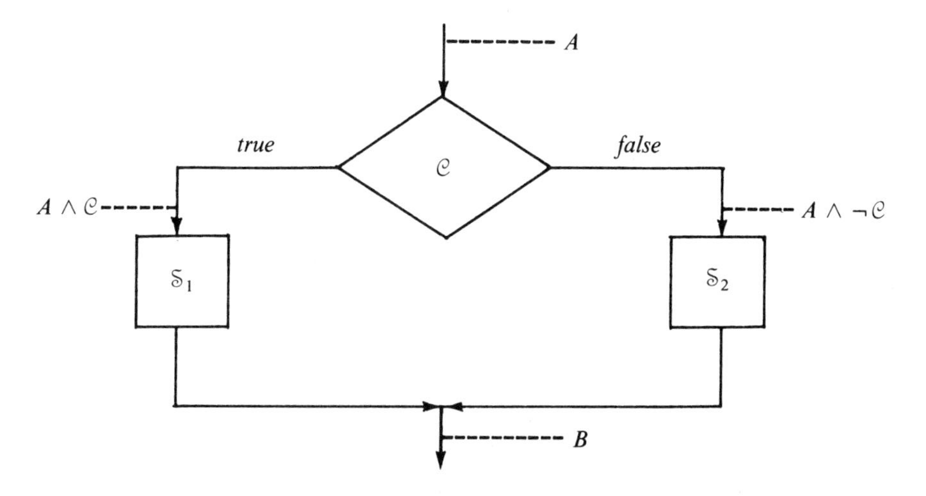

If we can derive both $\{A \wedge \mathcal{C}\}\mathcal{S}_1\{B\}$ and $\{A \wedge \neg\mathcal{C}\}\mathcal{S}_2\{B\}$, then we can assert that if the precondition A holds, then B will hold upon completion of execution of the conditional no matter which statement, $\mathcal{S}_1$ or $\mathcal{S}_2$, is selected for execution. So we can formulate the following proof rule, the *conditional statement rule*:

$$\frac{\{A \wedge \mathcal{C}\}\mathcal{S}_1\{B\}, \{A \wedge \neg\mathcal{C}\}\mathcal{S}_2\{B\}}{\{A\} \textbf{ if } \mathcal{C} \textbf{ then } \mathcal{S}_1 \textbf{ else } \mathcal{S}_2\{B\}} \qquad \text{(R5)}$$

We now turn to **while** statements. To justify our proof rule here, consider the flow diagram of **while** $\mathcal{C}$ **do** $\mathcal{S}$ on page 130, labelled with the appropriate pre- and postconditions:

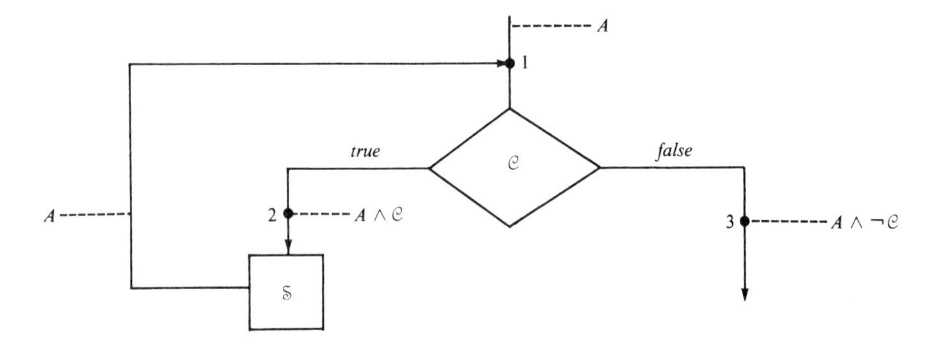

If A holds when we first enter, it is clear that $A \wedge \mathcal{C}$ will hold if we get to point 2. Thus, if we want to be assured that A again holds when we return to point 1, we would want to ensure that $\{A \wedge \mathcal{C}\}\mathcal{S}\{A\}$ was established for $\mathcal{S}$. In that case it is clear that A will hold not only when point 1 is reached for the first and second time, but also after an arbitrary number of cycles around the loop. Also $A \wedge \mathcal{C}$ holds whenever point 2 is reached. When the exit point 3 is reached, then not only A holds, but also $\neg\mathcal{C}$. So we obtain the **while** *statement rule*:

$$\frac{\{A \wedge \mathcal{C}\}\mathcal{S}\{A\}}{\{A\}\ \textbf{while}\ \mathcal{C}\ \textbf{do}\ \mathcal{S}\{A \wedge \neg\mathcal{C}\}} \qquad \text{(R6)}$$

Observe that (R6) establishes an *invariant property A* of the **while-do** loop. If A holds for the computation state initially, then A will hold for the computation state following each loop traversal. We also refer to A as a *loop invariant*.

We now illustrate the use of the proof rules with a few examples.

3 Example. We wish to prove the following program specification

$$\{X = 15\}\ \textbf{while}\ X > 10\ \textbf{do}\ X := pred(X)\ \{X = 10\}.$$

We list in sequence the steps of the derivation:

1. $(X \geqslant 10 \wedge X > 10) \supset (pred(X) \geqslant 10)$—an axiom.
2. $\{X \geqslant 10 \wedge X > 10\}X := pred(X)\{X \geqslant 10\}$—from (1), by (R1).
3. $\{X \geqslant 10\}$ **while** $X > 10$ **do** $X := pred(X)\{X \geqslant 10 \wedge \neg(X > 10)\}$—from (2), by (R6).
4. $(X = 15) \supset (X \geqslant 10)$—an axiom.
5. $\{X = 15\}$ **while** $X > 10$ **do** $X := pred(X)\{X \geqslant 10 \wedge \neg(X > 10)\}$—from (3) and (4), by (R3).
6. $(X \geqslant 10) \wedge \neg(X > 10) \supset (X = 10)$—an axiom.
7. $\{X = 15\}$**while** $X > 10$ **do** $X := pred(X)\{X = 10\}$—from (5) and (6), by (R2).

4 Example. We prove the program specification given in Example 2 (the integer division algorithm). Here are the steps of the derivation:

1. $(X \geqslant 0 \wedge Y > 0) \supset (X \geqslant 0 \wedge Y > 0)$—an axiom.
2. $\{X \geqslant 0 \wedge Y > 0\} V := 0 \{X \geqslant 0 \wedge Y > 0 \wedge V = 0\}$—from (1), by (R1).
3. $(X \geqslant 0 \wedge Y > 0 \wedge V = 0) \supset (X \geqslant 0 \wedge Y > 0 \wedge V = 0)$—an axiom.
4. $\{X \geqslant 0 \wedge Y > 0 \wedge V = 0\} W := X \{X \geqslant 0 \wedge Y > 0 \wedge V = 0 \wedge W = X\}$—from (3), by (R1).
5. $\{X \geqslant 0 \wedge Y > 0\} V := 0; W := X \{X \geqslant 0 \wedge Y > 0 \wedge V = 0 \wedge W = X\}$—from (2) and (4), by (R4).
6. $(X \geqslant 0 \wedge Y \geqslant 0 \wedge V = 0 \wedge W = X) \supset (X = V * Y + W \wedge 0 \leqslant W)$—an axiom.
7. $\{X \geqslant 0 \wedge Y > 0\} V := 0; W := X \{X = V * Y + W \wedge 0 \leqslant W\}$—from (5) and (6), by (R2).
8. $(X = V * Y + W \wedge 0 \leqslant W \wedge Y \leqslant W) \supset (X = (V + 1) * Y + W - Y \wedge 0 \leqslant W - Y)$—an axiom.
9. $\{X = V * Y + W \wedge 0 \leqslant W \wedge Y \leqslant W\} W := W - Y \{X = (V + 1) * Y + W \wedge 0 \leqslant W\}$—from (8), by (R1).
10. $(X = (V + 1) * Y + W \wedge 0 \leqslant W) \supset (X = (V + 1) * Y + W \wedge 0 \leqslant W)$—an axiom.
11. $\{X = (V + 1) * Y + W \wedge 0 \leqslant W\} V := V + 1 \{X = V * Y + W \wedge 0 \leqslant W\}$—from (10), by (R1).
12. $\{X = V * Y + W \wedge 0 \leqslant W \wedge Y \leqslant W\} W := W \dot{-} Y; V := V + 1 \{X = V * Y + W \wedge 0 \leqslant W\}$—from (9) and (11), by (R4).
13. $\{X = V * Y + W \wedge 0 \leqslant W\}$
    while $Y \leqslant W$ **do begin** $W := W \dot{-} Y;\ V := V + 1$ **end**
    $\{X = V * Y + W \wedge 0 \leqslant W \wedge \neg(Y \leqslant W)\}$—from (12), by (R6).
14. $\{X \geqslant 0 \wedge Y > 0\}$
    $V := 0; W := X;$
    while $Y \leqslant W$ **do begin** $W := W \dot{-} Y; V := V + 1$ **end** $\{X = V * Y + W \wedge 0 \leqslant W \wedge \neg(Y \leqslant W)\}$—from (7) and (13), by (R4).
15. $(X = V * Y + W \wedge 0 \leqslant W \wedge \neg(Y \leqslant W)) \supset (X = V * Y + W \wedge 0 \leqslant W \wedge W < Y)$—an axiom.
16. $\{X \geqslant 0 \wedge Y > 0\}$
    $V := 0; W := X;$
    while $Y \leqslant W$ **do begin** $W := W \dot{-} Y; V := V + 1$ **end**
    $\{X = V * Y + W \wedge 0 \leqslant W < Y\}$.

Two remarks are in order concerning the two previous examples. First, the formal proof of a relatively simple (indeed, trivial) program specification may be considerably longer than the program itself. In Example 3, we needed seven derivation steps, and in Example 4, 14 derivation steps. (For simple program properties that can be checked by inspection, formal proofs may seem hardly worth the effort on the basis of our exposition so far.)

Second, even with the simple examples given above, considerable ingenuity had to be used every time we introduced an axiom. Most difficult was the selection of an invariant property for the **while** statement in each example—$(X \geqslant 10)$ in Example 3, and $(X = V * Y + W \wedge 0 \leqslant W)$ in Example 4. Indeed, we could select the invariant property that made the proof go through in each case only because we knew intimately the function computed by the program. The moral is that one should develop a program and its formal specification together, using for example the methods spelled out by Alagić and Arbib (1978). With this methodology, formal program specifications and their proofs provide a valuable tool in the development of correct programs. Pre- and postconditions for program statements are part of the documentation supplied by a program designer (which is why we enclose them in braces { }, just as comments are when inserted in a PASCAL program).

On the more positive side of the preceding development, we note that our set of proof rules $\mathfrak{R} = \{(R1), \ldots, (R6)\}$ is not only sound, but also *complete*. This means that, given any program specification as in Definition 1, if it is true, then we can formally derive it using our set of axioms AX and proof rules $\mathfrak{R}$. The completeness of our logical system is established by Cook (1978).

Correct termination of algorithms

So far we have been concerned with correctness of results at a program's exit point, assuming the arguments satisfy the input condition. But this is merely one aspect of program correctness. It is equally important that the execution process yields results after only a finite number of steps. We now present a brief treatment of this problem, and show that loop invariants may play a crucial role in proving program termination.

Consider first the flow diagram of **while** $\mathcal{C}$ **do** $\mathcal{S}$, appropriately annotated, and let t be a term (which assumes a non-negative-integer value for every assignment of the variables):

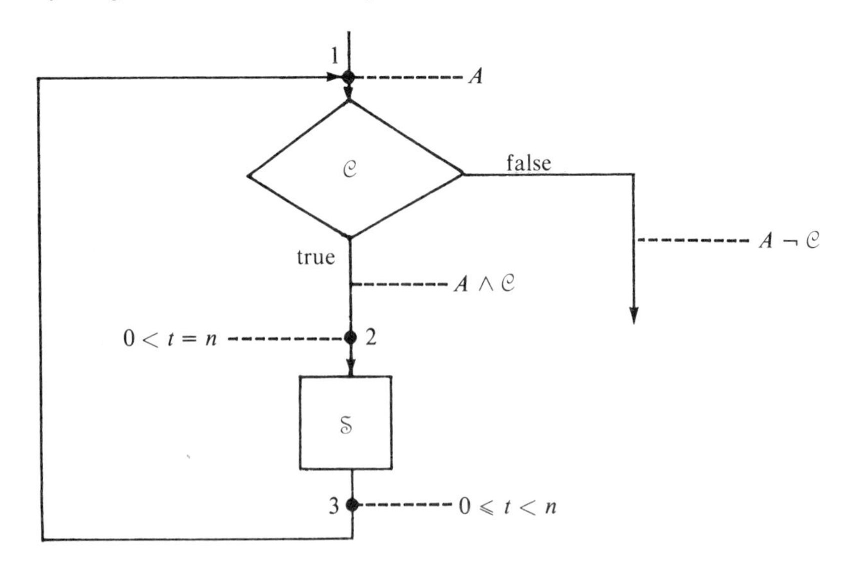

Suppose $A \wedge \mathcal{C} \supset (t > 0)$ and, for any fixed natural number n:

$$\{0 < t \wedge t = n\}\mathcal{S}\{0 \leqslant t < n\}$$

This partial correctness assertion says that exectuion of $\mathcal{S}$ decreases the value of t. This is indicated in the flow diagram by attaching the condition $(0 < t \wedge t = n)$ to point 2, and the condition $(0 \leqslant t < n)$ to point 3. Suppose that $\{A \wedge \mathcal{C}\}\mathcal{S}\{A\}$ has been established. Then the repetition process must be finite, since the value of t cannot be decreased infinitely often.

5 Example. Consider the division algorithm of Examples 2 and 4:

```
begin {(X ⩾ 0) ∧ (Y > 0)}
      V := 0; W := X;
      while W ⩾ Y do
      begin {W + Y > 0}
            W := W ∸ Y;
            V := V + 1
      end
      {W + Y > 0}
end
```

Here the desired term t is $W + Y$, since $W \geqslant 0$ and $Y > 0$ hold throughout any program execution. The value of $W + Y$ is reduced by every execution of the statement

$$\textbf{begin } W := W \dot{-} Y;\ V := V + 1 \textbf{ end}$$

and thus the termination is established.

When can we find a mistake in our program by showing that it does not terminate? Failing to come up with a suitable integer expression for proving termination does not, of course, permit us to conclude that our program does not terminate for some particular initial values of variables. Non-termination must be proven formally. There is no general algorithm for this, but suppose that we have a **while-do** loop (shown on page 134) for which we want to establish that $\{A\}$ **while** $\mathcal{C}$ **do** $\mathcal{S}\{A \wedge \neg\mathcal{C}\}$. If we suspect that such a loop does not terminate, we must prove that not only A, but also $A \wedge \mathcal{C}$ is an invariant of the loop. That is, we have to show that if $A \wedge \mathcal{C}$ holds at point 1, then point 3 will be reached, and $A \wedge \mathcal{C}$ will hold whenever that point is reached. If we establish this, then obviously the loop does not terminate.

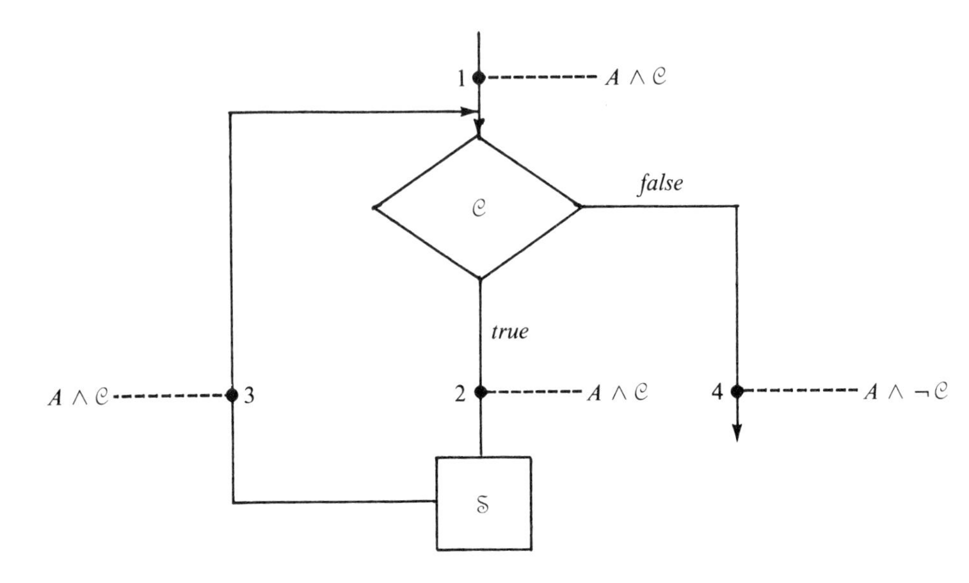

Further discussion of program termination, as a tool in the actual design of correct programs, can be found in Alagić and Arbib (1978). A more theoretical examination is given in deBakker (1980).

Exercises for Section 5.3

1. Prove each of the following program specifications:

   (a) $\{(X = m) \wedge (Y = n)\}$
   begin $Z := X; X := Y; Y := Z$ **end**
   $\{(Y = m) \wedge (X = n)\}$

   (b) $\{(Z = X * Y) \wedge \neg\, odd(X)\}$
   begin $Y := 2 * Y; X := X$ **div** 2 **end**
   $\{Z = X * Y\}$

   where $odd(X)$ is *true* if X is an odd number, and *false* if not.

2. Prove the following program specification:

   $\{(X = U + Y) \wedge (Y \leqslant V)\}$
   while $Y \neq V$ **do**
   begin $X := succ(X)$; $Y := succ(Y)$ **end**
   $\{X = U + V\}$

3. Prove the following program specification, where n is an arbitrary natural

number:

$$\{(X = n) \wedge (n \geqslant 0) \wedge (Y = 1) \wedge (Z \neq 0)\}$$
while $X > 0$ **do**
begin $Y := Y * Z; X := pred(X)$ **end**
$$\{(X = 0) \wedge (n \geqslant 0) \wedge (Y = Z^n) \wedge (Z \neq 0)\}$$

Use as a loop invariant $(X \geqslant 0) \wedge (n \geqslant 0) \wedge (Y = Z^{n-X}) \wedge (Z \neq 0)$.

4. Prove the following program specification, where n is an arbitrary natural number:

$$\{(X = n) \wedge (n \geqslant 0) \wedge (Y = 1)\}$$
while $X > 0$ **do**
begin $Y := Y + Y; X := pred(X)$ **end**
$$\{(X = 0) \wedge (n \geqslant 0) \wedge (Y = 2^n)\}$$

5. For all wff's A and B, and program statement $\mathbb{S}$, justify the following proof rule

$$\frac{\{A\}\mathbb{S}\{B\}}{\{A[Y/X]\}\mathbb{S}[Y/X]\{B[Y/X]\}}$$

provided Y is not free in B, and Y does not occur in $\mathbb{S}$. Exhibit an example of a derivation to show that, if the proviso "Y is not free in B and Y does not occur in $\mathbb{S}$" is omitted, the rule is not sound.

6. Let $\mathbb{S}$ be a program statement, and A and B wff's of first-order arithmetic. We say that A is a *weakest precondition* of $\mathbb{S}$ relative to postcondition B, in symbols $A = wp(\mathbb{S}, B)$, if:

(1) $\{A\}\mathbb{S}\{B\}$ is a true assertion.
(2) For all wff's A', if $\{A'\}\mathbb{S}\{B\}$ is a true assertion, then $A' \supset A$ is also true.

(a) Show that $wp(X := t, B)$ is $B[t/X]$, where t is an arbitrary term.
(b) Define, in a similar way, the notion of a *strongest postcondition* of $\mathbb{S}$ relative to precondition A, denoted by $sp(\mathbb{S}, A)$. What is $sp(X := t, A)$? Justify your answer informally.

7. We use the definition of "weakest precondition" given in the preceding exercise. Let $\mathbb{S}$ be the compound statement

begin $\mathbb{S}_1; \mathbb{S}_2; \ldots; \mathbb{S}_m$ **end**

Show that $wp(\mathbb{S}, B)$ for any postcondition B is

$$wp(\mathbb{S}_1, wp(\mathbb{S}_2, \ldots wp(\mathbb{S}_m, B) \ldots))$$

In particular, $wp(\mathbb{S}_1; \mathbb{S}_2, B) = wp(\mathbb{S}_1, wp(\mathbb{S}_2, B))$.

8. Use the definition of "weakest precondition" given in Exercise 6. What is

$wp($**if** $\mathbb{C}$ **then** $\mathbb{S}_1$ **else** $\mathbb{S}_2, B)$?

Justify your answer carefully.

9. (a) Use the following annotated flow diagram to provide a proof rule for **repeat** $\mathbb{S}$ **until** $\mathcal{C}$.

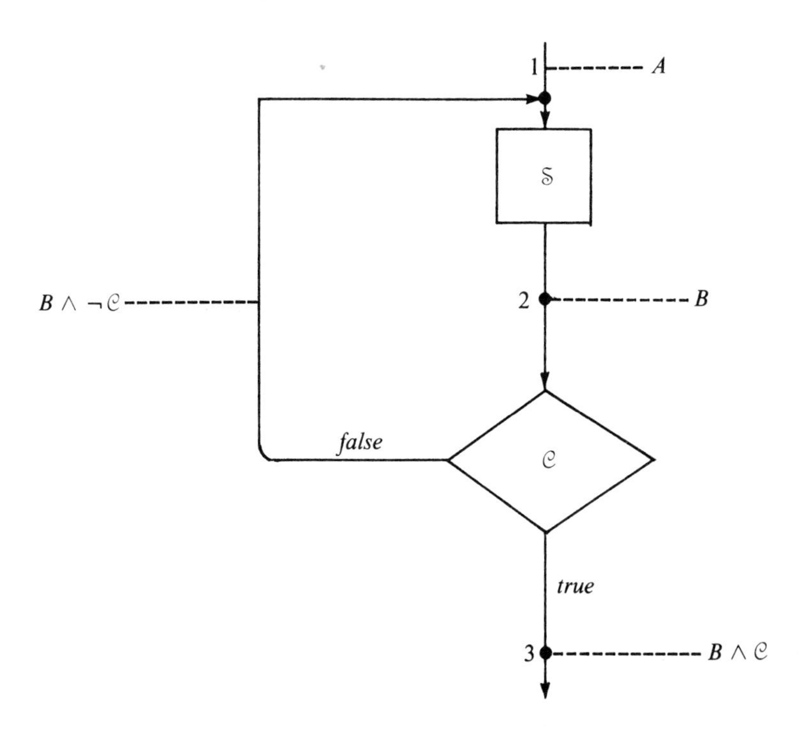

(b) Use the proof rules given in this section to verify that

begin $\mathbb{S}$; **while** $\neg\mathcal{C}$ **do** $\mathbb{S}$

satisfies the proof rule obtained in (a).

(c) Use the proof rule obtained in (a) in order to derive the following program specification:

$\{(X = U + Y) \wedge (Y < V)\}$
repeat begin $X := succ(X)$; $Y := succ(Y)$ **end**
until $Y = V$
$\{X = U + V\}$

10. Derive formally the following program specification

$\{(X > 0) \wedge (Y > 0)\}$
begin $Z := 0$; $U := X$;
  repeat begin $Z := Z + Y$; $U := U \dot{-} 1$ **end**
  until $U = 0$
end
$\{Z = X * Y\}$

Use Examples 3 and 4 as a guide for the derivation. (You need to establish first the proof rule for the **repeat-until** statement, as in Exercise 9.) An appropriate invariant for the **repeat-until** loop is $(Z + U * Y = X * Y)$.

11. Verify that for each of the preconditions given in Exercises 2, 3, and 4, the corresponding **while-do** loop terminates.

12. Prove termination of the multiplication algorithm given in Exercise 10. (*Hint*: Pick a quantity t that is reduced on each loop traversal.)

CHAPTER 6

The Recursion Theorem and Properties of Enumerations

Recursion is an important programming technique in computer science. We examined this notion initially in Section 5.2, where we also proved the First Recursion Theorem. Here we discuss a more general result, the Second Recursion Theorem (or more simply, the Recursion Theorem), perhaps the most fascinating result in elementary computability theory. The Recursion Theorem legitimizes self-reference in the descriptions of algorithms. Self-reference arguments add elegance and economy to the methods of computability theory, and guarantee that certain completely unexpected properties of any effective enumeration of algorithms must hold.

In Section 6.2 we invoke the Recursion Theorem, as well as several of the basic techniques presented in Chapter 4, to establish Rogers' Isomorphism Theorem. This theorem may be viewed as the theoretical justification of a model-independent development of computability theory.

6.1 The Recursion Theorem

Can there be a computable function $\varphi_e : \mathbf{N} \to \mathbf{N}$ which "describes itself":

$$\varphi_e(x) = e \text{ for all } x \in \mathbf{N}?$$

Such an index e encodes a **while**-program which outputs its own index, e, on every input. Apparently program e, i.e., $\mathbf{P}_e$, cannot encode this behavior directly; a program that includes statements for directly calculating e would seem to have an index larger than e.[1] But program e *can* contain a (sub)-program which, when called by the main program, *calculates* e indirectly and then outputs value e. This principle of indirect encoding is the underlying technique of the Recursion Theorem (also called the Second or Strong Recursion Theorem), and makes possible the apparently paradoxical algorithmic specification given above.

The Recursion Theorem, also due to Kleene, holds for any total computable function $f: \mathbf{N} \to \mathbf{N}$, i.e., for any *program-rewriting function* f, whereas the First Recursion Theorem holds only for program-rewriting functions induced by recursive programs (as defined in Section 5.2).

1 Theorem (The Recursion Theorem). *Let $f: \mathbf{N} \to \mathbf{N}$ be an arbitrary total computable function. Then for every arity j there exists an $n \in \mathbf{N}$ such that (in general, distinct) programs, $\mathbf{P}_n$ and $\mathbf{P}_{f(n)}$ compute the same j-ary function*:

$$\varphi_n^{(j)} = \varphi_{f(n)}^{(j)} .$$

[We often refer to an n for which $\varphi_n = \varphi_{f(n)}$ as a "fixed point" of f, even though this language is somewhat sloppy. It is not n that remains fixed as n changes to $f(n)$—usually $n \neq f(n)$—but rather it is the function $\varphi_n = \varphi_{f(n)}$ that stays the same.]

PROOF. With no loss of generality we prove the theorem for programs computing unary functions only, i.e., for $j = 1$. Recalling that $X := \Phi(Y, Z)$ is the macro for the interpreter of Chapter 3, we introduce the following program

$$\begin{aligned} \mathbf{P}_{g(i)} = \mathbf{begin}\ & X2 := \Phi(i, i); \\ & X1 := \Phi(X2, X1) \\ \mathbf{end}\ & \end{aligned}$$

where g is a total computable function, by the s-m-n Theorem (program form). Now consider the unary function computed by $\mathbf{P}_{g(i)}$, with variable $X1$ used for both input and output. Instruction $X2 := \Phi(i, i)$ terminates only if $\varphi_i(i)$ is defined, in which case $X2$ is assigned this $\varphi_i(i)$ as its value. But then the second line terminates just in case $\varphi_{\varphi_i(i)}(j)$ is defined, where j (unrelated to the arity mentioned in the theorem) is the initial value of $X1$, with $\varphi_{\varphi_i(i)}(j)$ as the final value of $X1$. We thus have

$$\varphi_{g(i)}(j) = \begin{cases} \varphi_{\varphi_i(i)}(j), & \text{if } \varphi_i(i)\downarrow; \\ \perp, & \text{otherwise.} \end{cases} \qquad (\$)$$

[1] For example, a program like

$$\mathbf{begin}\ X1 := 0; \underbrace{X1 := succ(X1); \ldots; X1 := succ(X1)}_{e \text{ times}}\ \mathbf{end}$$

has a numerical coding well in excess of e.

Readers should note that $g(i)$ is not the same as $\varphi_i(i)$, in general, since g is total but the map $i \mapsto \varphi_i(i)$ is not total. (Can you see why?)

Consider now an arbitrary total computable function $f: \mathbf{N} \to \mathbf{N}$ (the "program-rewriting function" of the theorem). The composite function $f \circ g$ is also computable and total, and therefore has an index, say m, among the computable functions from $\mathbf{N}$ to $\mathbf{N}$. Since $\varphi_m = f \circ g$ is total, $\varphi_m(m)$ is defined. Hence, substituting m for i in ($), we get

$$\varphi_{g(m)}(j) = \varphi_{\varphi_m(m)}(j), \text{ since } \varphi_m(m) \text{ is defined.}$$

But $\varphi_m = f \circ g$, and so we also have

$$\varphi_{g(m)}(j) = \varphi_{\varphi_m(m)}(j) = \varphi_{f \circ g(m)}(j),$$

for all j. Setting $n = g(m)$, we get the desired result. □

The Recursion Theorem is interesting because it tells us that any recursive equation of the form $\varphi_n = \varphi_{f(n)}$, where n is the "unknown" for a program-rewriting (i.e., total computable) function f, is satisfied by a **while**-program.

2 Examples. In each of the following cases, $f: \mathbf{N} \to \mathbf{N}$ is a program-rewriting function which is *not* induced by a recursive program in the sense of Section 5.2.

(1) f takes program n, i.e., $\mathbf{P}_n$, and replaces every occurrence of *succ* in $\mathbf{P}_n$'s text by *pred*, and every occurrence of *pred* by *succ*.

(2) f takes program n and looks up the first instruction I in it (which is either a "box" or a "diamond"). If I is of the form $Xi := succ(Xj)$, then f replaces I by

$$\textbf{begin } Xi := succ(Xj);\ Xi := succ(Xi) \textbf{ end}$$

If I is of the form $Xi := pred(Xj)$, then f replaces I by

$$\textbf{begin } Xi := pred(Xj);\ Xi := pred(Xi) \textbf{ end}$$

If I is neither of the two preceding forms, f inserts the text **begin** $X1 := 0$ **end** immediately after the program's first **begin**.

(3) f treats every **while**-program as a compound statment

$$\textbf{begin } \mathbb{S}_1; \mathbb{S}_2; \ldots; \mathbb{S}_m \textbf{ end}$$

and rewrites it backward, i.e., transforms it to

$$\textbf{begin } \mathbb{S}_m; \ldots; \mathbb{S}_2; \mathbb{S}_1 \textbf{ end}.$$

In each of the three preceding cases, there is a program n such that $\varphi_n = \varphi_{f(n)}$, by the Recursion Theorem.

If $f: \mathbf{N} \to \mathbf{N}$ is a program-rewriting function induced by a recursive program, then f has infinitely many fixed points, i.e., $\varphi_n = \varphi_{f(n)}$ for infinitely many n, because if program n satisfies $\varphi_n = \varphi_{f(n)}$, then program n'

obtained from program n by writing

$$\mathbf{P}_{n'} = \mathbf{begin}\ X1 := succ(X1);\ X1 := pred(X1);\ \mathbf{P}_n\ \mathbf{end}$$

is such that $\varphi_{n'} = \varphi_{f(n')}$ too (why?). If $f: \mathbf{N} \to \mathbf{N}$ is a program-rewriting function in general, then this trick of adding redundant instructions to program n satisfying $\varphi_n = \varphi_{f(n)}$ is not guaranteed to work anymore (Exercise 5). We nevertheless have the following result.

3 Proposition. *If $f: \mathbf{N} \to \mathbf{N}$ is a total computable function, then there are infinitely many n such that $\varphi_n^{(j)} = \varphi_{f(n)}^{(j)}$.*

PROOF. With no loss of generality let $j = 1$. We show that there must exist arbitrarily large n for which $\varphi_n = \varphi_{f(n)}$. Consider the first $k + 1$ computable functions, and choose a program l such that:

$$\varphi_l \neq \varphi_0,\ \varphi_l \neq \varphi_1,\ \ldots,\ \varphi_l \neq \varphi_k.$$

Now define a new total function $g: \mathbf{N} \to \mathbf{N}$ as follows:

$$g(x) = \begin{cases} l, & \text{if } x \leqslant k; \\ f(x), & \text{if } x > k. \end{cases}$$

By Church's Thesis, g is computable and therefore has a fixed point in the sense of the Recursion Theorem, say n. If $n \leqslant k$ then $\varphi_{g(n)} = \varphi_l \neq \varphi_n$, which contradicts the fact that n is a fixed point for g. Hence $n > k$, so that $f(n) = g(n)$ and n is a fixed point for f. Moreover n is arbitrarily large, since k was chosen arbitrarily. □

There are many important applications of the Recursion Theorem. In particular, it can be used to justify (rather unexpected) recursive definitions of computable functions.

4 Example. There is a **while**-program that outputs its own description, i.e., there is a computable function $\varphi_e: \mathbf{N} \to \mathbf{N}$ such that $\varphi_e(b) = e$, for all $b \in \mathbf{N}$.

Define the program-rewriting function $f: \mathbf{N} \to \mathbf{N}$ which takes an arbitrary $\mathbf{P}_i$ to $\mathbf{P}_{f(i)}$:

$$\mathbf{P}_{f(i)} = \mathbf{begin}\ X1 := i\ \mathbf{end}.$$

f has a fixed point, say e. Hence,

$$\varphi_e(b) = \varphi_{f(e)}(b) = e,$$

for all $b \in \mathbf{N}$.

SELF-REFERENCE

Before examining other applications of the Recursion Theorem, let us try to isolate the self-referential aspects of the result. [The discussion which follows is adapted from Rogers (1967), pp. 199–200.]

The proof of the Recursion Theorem involves the construction of a total computable function g such that

$$\varphi_{g(m)}(j) = \varphi_{\varphi_m(m)}(j) \text{ for all } j \in \mathbf{N}.$$

That is, the *instruction* $g(m)$ says:

> "Compute program m on argument m. If and when this computation terminates, interpret the output value as a program, and apply *that* program to input j."

This is the meaning of $g(m)$ for any m. Self-reference is introduced when m involves an encoding of g itself. More specifically, if m is a code for $f \circ g$, then $g(m)$ says:

> "Compute $f(g(m))$. Then apply program $f(g(m))$ to input j."

That is to say, program $g(m)$ first calculates $f(g(m))$, and then executes $f(g(m))$; program $f(g(m))$, on the other hand, merely executes $f(g(m))$. Both programs *do* $f(g(m))$, so

$$\varphi_{g(m)} = \varphi_{f(g(m))},$$

but program $g(m)$ takes longer to calculate a value on input j, since it must first compute (a code for) program $f(g(m))$.

Another way to view the instruction set $g(m)$, where m is an index for $f \circ g$, is to permit an anthropomorphic assertion in which $g(m)$ *describes* its own behavior. Instruction $g(m)$ says:

> "I apply function f to myself. When this computation terminates, I apply the result to input j."

From this perspective, self-reference is explicit. Indeed once it is recognized that such references are allowed in program descriptions, then applying the Recursion Theorem becomes a simple and, indeed, almost trivial exercise. For example let us reconsider Example 4. In that example we showed formally that there is an e such that $\varphi_e(x) = e$. If we allow anthropomorphic self-reference to replace an explicit fixed-point argument—which we must be able to supply on demand—then the following self-referential specification is valid:

> "I calculate my own number, and output this value on all inputs."

This statement justifies the specification:

$$\varphi_e(j) = e \text{ for all } j \in \mathbf{N}.$$

5 Example. There is a **while**-program which is defined only on its own description. We can define such a program easily, using a self-referential description:

$$\varphi_{e^*}(x) = \begin{cases} 0, & \text{if } x = e^*; \\ \bot, & \text{otherwise.} \end{cases}$$

In other words, program e^* says: "I compare my own number to the input, x. If the two are equal, I output 0. Otherwise I diverge". To prove the existence of program e^*, define the program-rewriting function $f: \mathbf{N} \to \mathbf{N}$ by setting:

$$\mathbf{P}_{f(e)} = \mathbf{begin\ while}\ X1 \neq e\ \mathbf{do}\ X1 := X1;\ X1 := 0\ \mathbf{end}$$

We have

$$\varphi_{f(e)}(x) = \begin{cases} 0, & \text{if } x = e; \\ \bot, & \text{otherwise.} \end{cases}$$

The function f has a fixed point, say e^*, and so

$$\varphi_{f(e^*)}(x) = \varphi_{e^*}(x) = \begin{cases} 0, & \text{if } x = e^*; \\ \bot, & \text{otherwise.} \end{cases}$$

6 Example. There exists an integer i^* such that for all j,

$$\varphi_{i^*}(j) = \varphi_j(i^*).$$

Again, if specifications involving self-reference are credible, then this result is eminently believable: "I look at my own number and my argument. I interpret my argument as a program, and I run that program on my number". We establish the existence of i^* by defining the program-rewriting function $f: \mathbf{N} \to \mathbf{N}$ by setting:

$$\mathbf{P}_{f(i)} = \mathbf{begin}\ X1 := \Phi(X1, i)\ \mathbf{end}$$

We clearly have

$$\varphi_{f(i)}(j) = \begin{cases} \varphi_j(i), & \text{if } \varphi_j(i)\downarrow; \\ \bot, & \text{otherwise.} \end{cases}$$

By the Recursion Theorem, f has a fixed point i^*:

$$\varphi_{i^*}(j) = \varphi_{f(i^*)}(j) = \begin{cases} \varphi_j(i^*), & \text{if } \varphi_j(i^*)\downarrow; \\ \bot, & \text{otherwise;} \end{cases}$$

which can be simplified to read $\varphi_{i^*}(j) = \varphi_j(i^*)$.

7 Example. Consider the sequence

$$p_i(x) = \begin{cases} 0, & \text{if } x < i; \\ 2^{(x-i)}, & \text{otherwise.} \end{cases}$$

The p_i's have the property that they form an infinite decreasing sequence of increasing functions. That is,

(a) for all i and x, $p_i(x) \leqslant p_i(x+1)$,
(b) for all i and x, $p_{i+1}(x) \leqslant p_i(x)$,
(c) for any i, and all but finitely many values of x, $2 * p_{i+1}(x) \leqslant p_i(x)$.

Using the Recursion Theorem we can generalize this construction to build infinite descending sequences of computable functions with almost any conceivable computable property. For example, suppose we wish to build a sequence of total computable functions $\{g_i\}_{i \in \mathbf{N}}$, such that

(a) for all i, $g_i(x) \geqslant b(x)$ for all but finitely many x, where $b(x)$ is any fixed total computable function;
(b) for any i, and for all but finitely many x, $g_{i+1} \circ g_{i+1}(x) \leqslant g_i(x)$.

We proceed as follows. Let us agree that we seek an index $t^* \in \mathbf{N}$ such that $\{\varphi_{s_1^1(t^*,i)}\}_{i \in \mathbf{N}}$ is our desired sequence, with $\varphi_{s_1^1(t^*,i)}$ identified as g_i. We write

$$\theta(t,i,x) = \begin{cases} 0, & \text{if } x < i \text{ or there is some } n < x \text{ such that } \varphi_{s_1^1(t,0)}(n) \text{ fails to halt after } x \text{ steps;} \\ \varphi_{s_1^1(t,i+1)} \circ \varphi_{s_1^1(t,i+1)}(x) + b(x), & \text{otherwise.} \end{cases}$$

Now θ is computable by the following argument. On inputs t, i, and x, check first to see if $i < x$, and if so output 0. Otherwise run $\mathbf{P}_{s_1^1(t,0)}$—the program that computes g_0—for x steps on each of the inputs $0, \ldots, x-1$. If one of the computations fails to halt within the allotted x steps, output 0. Otherwise, apply $\mathbf{P}_{s_1^1(t,i+1)}$ to the output of $\mathbf{P}_{s_1^1(t,i+1)}$, add $b(x)$, and output the result (if there is one). Hence θ is computable, say $\theta(t,i,x) = \varphi_e(t,i,x)$ $= \varphi_{s_1^1(e,t)}(i,x)$. Setting $s_1^1(e,t)$ to $f(t)$, we get $\theta(t,i,x) = \varphi_{f(t)}(i,x)$. f is total computable, so it has a fixed point which we call t^*, yielding

$$\varphi_{s_1^1(t^*,i)}(x) = \varphi_{t^*}(i,x) = \begin{cases} 0, & \text{if } x < i, \text{ or there is an } n < i \text{ such that } \varphi_{s_1^1(t^*,0)}(n) \text{ fails to halt after } x \text{ steps;} \\ \varphi_{s_1^1(t^*,i+1)} \circ \varphi_{s_1^1(t^*,i+1)}(x) + b(x), & \text{otherwise.} \end{cases}$$

Let us rewrite this as

$$g_i(x) = \begin{cases} 0, & \text{if } x < i \text{ or there is an } n < i \text{ such that } g_0(n) \text{ fails to halt after } x \text{ steps;} \quad (1) \\ g_{i+1} \circ g_{i+1}(x) + b(x), & \text{otherwise.} \quad (2) \end{cases}$$

To complete the proof, we must show that the sequence $\{g_i\}_{i \in \mathbf{N}}$ has the desired properties. We claim that it is sufficient to show that g_i is total for every i. Indeed if this is so, then g_0 is total. Hence, for any fixed i, the set

$$\{x \mid x < i \text{ or there is an } n < i \text{ such that } g_0(n) \text{ fails to halt after } x \text{ steps}\}$$

is finite. Hence, for all but finitely many x, we have $g_i(x) = g_{i+1} \circ g_{i+1}(x) + b(x) \geqslant b(x)$—and we are done.

So let $m \in \mathbf{N}$ be the smallest value for which g_m is nontotal. If $m = 0$, then $g_0(j)$ is undefined for some j. When $i > j$, $g_i \equiv 0$, since clause (1) always applies in the definition of g_i. Hence, for all r, $0 \leqslant r < i$, g_r is total, since g_r's values are either 0 by clause (1), or are inherited from g_{r+1}, and, ultimately, from the total function g_i, by clause (2).

Now suppose that $m \neq 0$ is the smallest value for which g_m is not total. Then by assumption g_{m-1} *is* total. Now if $g_m(j)$ is undefined, then j must satisfy clause (2), because if j satisfied clause (1) the value of $g_m(j)$ would be 0 (and therefore defined!). Thus clause (1) must not apply to $g_m(j)$, meaning that

(a) $j \geqslant m$,
(b) $g_0(n)$ must converge within j steps for all $n < m$.

But then

(a) $j \geqslant m - 1$,
(b) $g_0(n)$ must converge within j steps for all $n < m - 1$,

implying that $g_{m-1}(j)$ is defined to be equal to $g_m \circ g_m(j) + b(j)$. But then $g_{m-1}(j)$ is divergent also, contradicting the minimality of m. Thus for all m, g_m is total, and this completes the proof.

Exercises for Section 6.1

1. Give an example of a total computable function with no actual fixed point. That is, find an f such that $f(i) \neq i$ for all $i \in \mathbf{N}$.

2. What is a fixed point, in the sense of the Recursion Theorem, for the constant function $f(x) \equiv a$ for some $a \in \mathbf{N}$?

3. Does there exist an e such that

$$\varphi_e(x) = \begin{cases} 1, & \text{if } x \neq e; \\ \bot, & \text{otherwise.} \end{cases}$$

Justify your answer.

4. Consider Example 2. In each of the cases defined, (1), (2), and (3), find two fixed points n_1 and n_2 for the equation $\varphi_n = \varphi_{f(n)}$ such that $\varphi_{n_1} \neq \varphi_{n_2}$. Justify your answers.

5. Give a program-rewriting function $f: \mathbf{N} \to \mathbf{N}$, and a fixed point n for it, such that if we define program n' to be:

$$\mathbf{P}_{n'} = \textbf{begin } X1 := succ(X1); X1 := pred(X1); \mathbf{P}_n \textbf{ end}$$

then n' is not a fixed point for f, i.e., $\varphi_{n'} \neq \varphi_{f(n')}$. Note that if f were a program-rewriting function induced by a recursive program, then n' would be a fixed point. (*Hint*: Consider the first program-rewriting function in Example 2.)

6. Discuss in detail the computational status of the following partial function θ:

$$\theta(x, y) = \begin{cases} \varphi_{\varphi_{\varphi_x(x)}(x)}(y), & \text{if } \varphi_x(x)\downarrow \text{ and } \varphi_{\varphi_x(x)}(x)\downarrow; \\ \perp, & \text{otherwise.} \end{cases}$$

7. Show that the Recursion Theorem does not hold when restricted to total functions. That is, show that the following statement is not a theorem: "Suppose $f: \mathbf{N} \to \mathbf{N}$ is a total computable function such that if φ_x is total then $\varphi_{f(x)}$ is total. Then f has a fixed point n such that $\varphi_n = \varphi_{f(n)}$ and φ_n is total".

8. Does there exist an e such that

$$\varphi_e(x) = \begin{cases} 1, & \text{if } \varphi_e(k)\uparrow \text{ for some } k < x; \\ \perp, & \text{otherwise.} \end{cases}$$

Justify your answer.

9. (Rogers) Let f be a total computation function. Argue that the function $\theta(x, y) = \varphi_{f \circ s_1^1(x,x)}(y)$ is computable, and therefore has an index e, $\theta = \varphi_e^{(2)}$. Set $n = s_1^1(e, e)$. Prove that $\varphi_n = \varphi_{f(n)}$. (This presents an alternative proof for the Recursion Theorem.)

10. (a) Using the proof of the Recursion Theorem given in the text as a guide, prove the following extended Recursion Theorem:

    Let $g: \mathbf{N}^2 \to \mathbf{N}$ be a total computable function. Then there exists a total computable function $f: \mathbf{N} \to \mathbf{N}$ such that $\varphi_{f(n)} = \varphi_{g(n, f(n))}$ for all n.

    (b) Use part (a) to prove the existence of a total computable function f such that for all n,

$$\varphi_{f(n)}(x) = \begin{cases} 1, & \text{if } x = f(n) + n; \\ \perp, & \text{otherwise.} \end{cases}$$

11. Prove that there exists a total computable function of one variable $f: \mathbf{N} \to \mathbf{N}$ such that $\varphi_{f(n)} = \varphi_{f \circ f(n)}$.

12. Use the Recursion Theorem to prove the unsolvability of the Totality Problem. That is, show that the set $T = \{n \mid \varphi_n \text{ is total}\}$ is not solvable. (*Hint*: Assume the contrary and consider the self-referential specification

$$\varphi_m(x) = \begin{cases} \perp, & \text{if } m \in T; \\ 1, & \text{otherwise.}) \end{cases}$$

13. (a) Use the Recursion Theorem to prove that the set of indices $\{n \mid \varphi_n(n)\uparrow\}$ is unsolvable. That is, it is impossible to verify algorithmically whether an arbitrary **while**-program diverges on its own index.
    (b) Reduce the solvability of the set $\{n \mid \varphi_n(n)\uparrow\}$ to the solvability of $\{(i, j) \mid \varphi_i(\varphi_j(i))\uparrow\}$ to show that the latter set is not decidable.
    (c) By a technique of your own choosing, prove that the set $\{n \mid \varphi_n(n^2 + n + 1)\uparrow\}$ is not solvable.

14. Use the Recursion Theorem to show that the set $\{n \mid \varphi_n \equiv 0\}$ is not solvable; i.e., that it is impossible to verify algorithmically that an arbitrary **while**-program computes the function which identically equals 0.

15. Construct a sequence of total computable functions $\{h_i\}_{i \in \mathbf{N}}$ such that:

(1) For each i and for all but finitely many x, $h_i(x) \geqslant b(x)$, where b is some fixed but arbitrary total computable function.
(2) For all i and all but finitely many x,
$$h_i(x) > max\{h_{i+1}(y) \mid y \leqslant x\}.$$

6.2 Model-Independent Properties of Enumerations

Although many of our proofs to date involve explicit program constructions, these can apparently be eliminated by appropriately appealing to Church's Thesis, the s-m-n Theorem, and the Enumeration Theorem. We wish to show that with these theorems we can indeed discard explicit program constructions and make computability theory fully model independent.

Acceptable programming systems

The following definition, introduced by Rogers, is commonly found in the literature (in various equivalent formulations).

1 Definition. Let $A_0, A_1, A_2, \ldots$ be an effective enumeration of all algorithms in some algorithm-specification system. We assume that for every arity $j \in \mathbf{N}$, we can associate with this enumeration an effectively indexed sequence of j-ary functions $\alpha_0^{(j)}, \alpha_1^{(j)}, \alpha_2^{(j)}, \ldots,$ where algorithm A_i computes the j-ary function $\alpha_i^{(j)}$. We shall say that the sequence $A_0, A_1, A_2, \ldots$ is an *acceptable programming system*, and the corresponding indexing of the j-ary functions they compute is an *acceptable numbering* if:

(1) Every computable function on $\mathbf{N}$ is in the set $\{\alpha_i^{(j)} \mid i, j \in \mathbf{N}\}$; that is, the programming system is *complete* (as defined in Section 1.1).
(2) For every k, the sequence of k-ary functions $\alpha_0^{(k)}, \alpha_1^{(k)}, \alpha_2^{(k)}, \ldots$ has a universal function $\alpha_u^{(k+1)} \in \{\alpha_i^{(j)} \mid i, j \in \mathbf{N}\}$; that is, the programming system satisfies the *universal function property*.
(3) For every m and n, there is a total computable s_n^m function in the list $\alpha_0^{(m+1)}, \alpha_1^{(m+1)}, \alpha_2^{(m+1)}, \ldots$ such that $\alpha_{s_n^m(i, y_1, \ldots, y_m)}^{(n)}(z_1, \ldots, z_n) = \alpha_i^{(m+n)}(y_1, \ldots, y_m, z_1, \ldots, z_n)$; that is, the programming system satisfies the *s-m-n property*.

Note that there is a difference in status between conditions (1), (2), and (3). As discussed in Chapter 1, the completeness of an algorithm-specification system cannot be proved, and is a credible claim only from lack of evidence to the contrary (Church's Thesis). On the other hand, conditions (2) and (3) were proved for the system of algorithms based on **while**-programs. We therefore alter (1) by taking "completeness" to mean "computing all functions computable by **while**-programs".

We do not require that an acceptable programming system explicitly satisfy the Recursion Theorem, since it is a consequence of the three defining properties of an acceptable programming system.

We now show that any two acceptable programming systems are effectively intertranslatable. Let system $A_0, A_1, A_2, \ldots$ compute the set of functions $\{\alpha_i^{(j)} \mid i, j \in \mathbf{N}\}$, and let system $B_0, B_1, B_2, \ldots$ compute $\{\beta_i^{(j)} \mid i, j \in \mathbf{N}\}$. (The A_i's and the B_i's are not necessarily based on different models of computation. They can be two distinct effective enumerations of the same system, such as **while**-programs.)

2 Proposition. *Let $A_0, A_1, A_2, \ldots$ and $B_0, B_1, B_2, \ldots$ be acceptable programming systems. For each j there is a total computable function $t: \mathbf{N} \to \mathbf{N}$ which translates the A_i's to the B_i's; that is, $\alpha_i^{(j)} = \beta_{t(i)}^{(j)}$.*

Proof. With no loss of generality, let $j = 1$. Let $\alpha_u^{(2)}: \mathbf{N}^2 \to \mathbf{N}$ be a universal function for the sequence $\alpha_0, \alpha_1, \alpha_2, \ldots$. By the completeness property, there is an index k in the second system $B_0, B_1, B_2, \ldots$ such that $\alpha_u^{(2)} = \beta_k^{(2)}$. We now have:

$$\alpha_i(x) = \alpha_u^{(2)}(i, x) = \beta_k^{(2)}(i, x) = \beta_{s_1^1(k,i)}(x),$$

for all i. The last equality follows from the fact that $B_0, B_1, B_2, \ldots$ satisfy the s-m-n property. Setting $t(i) = s_1^1(k, i)$, as a function of i, we obtain the desired translation function. □

The function t in the above proposition is not necessarily one-to-one or onto. Using the Recursion Theorem, we can prove a stronger result in which t is one-to-one.

3 Proposition. *Assuming the hypothesis of Proposition 2, for each j there is a one-to-one total computable function $t: \mathbf{N} \to \mathbf{N}$ which translates the A_i's to the B_i's, i.e., $\alpha_i^{(j)} = \beta_{t(i)}^{(j)}$.*

Proof. Exercise 3. □

We show next that this result can be strengthened further by making the function t both one-to-one and onto.

Rogers' isomorphism theorem

In this subsection we present a theorem, due to Rogers (1958), which establishes an "effective isomorphism" between any two acceptable programming systems. This implies there is an abstract approach to computability theory: Starting from an acceptable numbering of computable functions, the entire theory can be developed without explicit reference to a specific model of computation.

4 Theorem (Rogers' Isomorphism Theorem). *Let $A_0, A_1, A_2, \ldots$ and $B_0, B_1, B_2, \ldots$ be acceptable programming systems. For each j there is a one-to-one, onto, total computable function $t: \mathbf{N} \to \mathbf{N}$ which translates the A_i's to the B_i's such that $\alpha_i^{(j)} = \beta_{t(i)}^{(j)}$.*

For the proof of the theorem, we need a lemma. With no loss of generality we restrict our attention to unary functions, i.e., $j = 1$ in the statement of the theorem. Let us call a finite sequence of ordered pairs of indices $(x_1, y_1), (x_2, y_2), \ldots, (x_n, y_n)$, a *finite correspondence* (between the A_i's and B_i's) if:

(1) $i \neq j$ implies $x_i \neq x_j$ and $y_i \neq y_j$,
(2) $\alpha_{x_i} = \beta_{y_i}$,

for all $1 \leqslant i, j \leqslant n$.

5 Lemma. *There is an effective procedure such that if $(x_1, y_1), \ldots, (x_n, y_n)$ is a finite correspondence and if $x' \notin \{x_1, \ldots, x_n\}$, then we can find a y' such that $(x_1, y_1), \ldots, (x_n, y_n), (x', y')$ is a finite correspondence.*

PROOF. Let t be the one-to-one translation function defined in Proposition 3, which establishes that for all x, $\alpha_x = \beta_{t(x)}$.

Given $x' \notin \{x_1, \ldots, x_n\}$, consider $t(x')$ and see if $t(x') \neq y_i$ for all i, $1 \leqslant i \leqslant n$; if so, set $y' = t(x')$. If not, and $t(x') = y_{i_1}$ for some i_1, consider $t(x_{i_1})$ and see if $t(x_{i_1}) \neq y_i$ for all i, $1 \leqslant i \leqslant n$; if so, set $y' = t(x_{i_1})$. If not, and $t(x_{i_1}) = y_{i_2}$ for some i_2, consider $t(x_{i_2})$ and see if $t(x_{i_2}) \neq y_i$ for all i, $1 \leqslant i \leqslant n$; etc. Since t is one-to-one and since the given correspondence is finite, this procedure must end with the determination of an appropriate y' such that $y' \notin \{y, \ldots, y_n\}$ and $\alpha_{x'} = \beta_{y'}$. Note that the one-to-one property of t insures that $i_1 \neq i_2 \neq \cdots$. □

PROOF OF THEOREM 4. With no loss of generality, let $j = 1$. Let t be a one-to-one translation from the A_i's into the B_i's, and u a one-to-one translation from the B_i's to the A_i's; $\alpha_x = \beta_{t(x)}$ and $\beta_y = \alpha_{u(y)}$ for all x, y. By the symmetry of the conditions, note that the lemma also holds with "$x' \notin \{x_1, \ldots, x_n\}$" replaced by "$y' \notin \{y_1, \ldots, y_n\}$" and "$y'$ such that" replaced by "x' such that" in its statement; for this form, the proof uses u instead of t.

We now construct an effective isomorphism (one-to-one, onto, total computable function) between the A_i's and the B_i's. We give an effective procedure for listing pairs of integers. The construction is arranged so that at each stage the ordered pairs already listed constitute a finite correspondence and, furthermore, so that every natural number occurs eventually as the first member of some ordered pair in the list *and* every natural number occurs eventually as the second member of some ordered pair in the list. The entire collection of ordered pairs listed thus constitutes an effective isomorphism between the A_i's and the B_i's.

We define the procedure in stages.

Stage 0. Take $(0, t(0))$ as the first ordered pair.

Stage 1. See whether $t(0) = 0$. If so, proceed to stage 2. If not, use the lemma to list a new ordered pair whose second member is 0.

⋮

Stage $2n$. See whether n occurs as the first member of some pair already listed. If so, go to stage $2n + 1$. If not, use the lemma to list a new ordered pair whose first member is n.

Stage $2n + 1$. See whether n occurs as the second member of some pair already listed. If so, go to stage $2n + 2$. If not, use the lemma to list a new ordered pair whose second member is n.

⋮

The verification of this procedure is straightforward. □

We may view the preceding theorem as the counterpart in computability theory of the Cantor–Bernstein theorem in set theory, which shows that if there is an injection from a set A to a set B, and an injection from B to A, then there is a bijection between A and B.

Exercises for Section 6.2

1. Let $A_0, A_1, A_2, \ldots$ and $B_0, B_1, B_2, \ldots$ be two acceptable programming systems. Show that there must be a unary function that has the same index in both systems.

2. (Rogers) Let $A_0, A_1, A_2, \ldots$ and $B_0, B_1, B_2, \ldots$ be two acceptable programming systems. By Proposition 2, there are total computable functions $t : \mathbf{N} \to \mathbf{N}$ and $u : \mathbf{N} \to \mathbf{N}$ such that $\alpha_i = \beta_{t(i)}$ and $\beta_i = \alpha_{u(i)}$. Using the fact that the Recursion Theorem holds for the unary functions $\{\alpha_i \mid i \in \mathbf{N}\}$ computed by the A_i's, prove that it also holds for $\{\beta_i \mid i \in \mathbf{N}\}$ computed by the B_i's, i.e., show that for any total computable function $f : \mathbf{N} \to \mathbf{N}$, there is an n such that $\beta_n = \beta_{f(n)}$. (*Hint*: First obtain an m such that $\alpha_m = \alpha_{u \circ f \circ t(m)}$.)

3. Consider the translation function t of Proposition 2.

   (a) Show that if the programming system $B_0, B_1, B_2, \ldots$ is our standard enumeration of all **while**-programs, then t is one-to-one. (*Hint*: Our s_n^m functions in Section 4.2 are one-to-one.)

   (b) Let $B_0, B_1, B_2, \ldots$ be an arbitrary acceptable programming system. Let s_n^m denote the *s-m-n* functions of system $B_0, B_1, B_2, \ldots$ which are not necessarily one-to-one.

(i) Use the Recursion Theorem (relative to system $B_0, B_1, B_2, \ldots$) to prove that for every total computable function $f: \mathbf{N} \to \mathbf{N}$, there is a fixed index i such that:

$$\beta_i(x, y) = \begin{cases} 0, & \text{if there is an } n < x \text{ such that} \\ & s_1^1(i, n) = s_1^1(i, x); \\ 1, & \text{if for all } n < x,\ s_1^1(i, n) \neq s_1^1(i, x), \\ & \text{and there is an } n \text{ such that } x < n \leqslant y \text{ and} \\ & s_1^1(i, x) = s_1^1(i, n); \\ \beta_{f(x)}(y), & \text{otherwise.} \end{cases}$$

Use this definition of index i to show that $s_1^1(i, x)$ is one-to-one as a function of x, satisfying, in addition,

$$\beta_{s_1^1(i,x)}(y) = \beta_i(x, y) = \beta_{f(x)}(y).$$

(ii) Use part (i) to show there exists a one-to-one translation (total computable function) $\hat{t}$ from system $A_0, A_1, A_2, \ldots$ to system $B_0, B_1, B_2, \ldots$ such that $\alpha_i = \beta_{\hat{t}(i)}$.

CHAPTER 7

Computable Properties of Sets (Part 1)

In preceding chapters we examined the class of computable functions in detail. We set up a coding system for these functions, using the natural numbers to represent the programs which compute these functions. Conveniently, the natural numbers also serve as the domain and codomain for functions in the class. Using this machinery, we were able to identify an important role for self-reference in our theory, and we were able to present natural examples of functions which are *not* computable, i.e., not calculable by an algorithm.

In this chapter we develop an analogous theory of computability for sets of natural numbers. In this theory two classes of sets, the recursive sets and recursively enumerable sets, are the main objects of study. Recursive sets are sets which are decidable, i.e., there is an algorithm for telling whether or not an element belongs to a recursive set. If a set is recursively enumerable, on the other hand, then we know only that a method exists for listing all its elements in some order.

These two classes do not coincide. Every recursive set is recursively enumerable, but there are recursively enumerable sets which are not recursive, and these latter sets play a crucial role in computability theory. Indeed, the distinction between these two concepts is older than computability theory itself, and has its roots in the Austrian mathematician Kurt Gödel's result establishing the incompleteness of formal systems of arithmetic in 1931. We devote Section 7.3 to a discussion of that famous theorem.

7.1 Recursive and Recursively Enumerable Sets

Is there a way, compatible with the previous results, in which we may talk about "computable" and perhaps also "partially computable" sets? Our study of the computable functions has laid the groundwork for our study of such objects.

Recursive sets

The simplest way to specify a set $A \subset \mathbf{N}^k$ by a function is in terms of its *characteristic function* $\chi_A : \mathbf{N}^k \to \{0, 1\}$ defined by

$$\chi_A(x) = \begin{cases} 1, & \text{if } x \in A; \\ 0, & \text{otherwise.} \end{cases}$$

(Even though the characteristic function of an arbitrary set A is always total, we follow the usual notation of denoting it by the Greek letter χ, subscripted with A.)

1 Definition. A set $A \subset \mathbf{N}^k$ is *decidable* or *solvable* or *recursive* if its characteristic function χ_A is a total computable function. This means that it is always possible to decide whether or not an arbitrary $x \in \mathbf{N}^k$ belongs to the set A, since there is an always-converging **while**-program to evaluate χ_A.

The fact that we have more than one term for the same concept—"decidable set", "solvable set", "recursive set", and "computable set"—is historically related to the different alternative characterizations of computability which we discuss in Chapter 9.

2 Examples. (i) The set of all prime natural numbers is a recursive set. High school mathematics gives us several different procedures to determine whether an arbitrary natural number is a prime.

(ii) The set of all pairs (a, b) of natural numbers, such that b is the square of a, is a recursive set: To decide whether an arbitrary $(a, b) \in \mathbf{N}^2$ is a member of the set, we compute $a * a$ and compare it to b.

By rephrasing decision problems as problems of set membership, we can easily exhibit interesting sets that are not recursive.

3 Examples. (i) The set $\{i \mid \varphi_i : \mathbf{N} \to \mathbf{N} \text{ is total}\}$ is not recursive, since its characteristic function *total* cannot be effectively computed, as established by Theorem 4 of Section 4.3.

(ii) The set $\{(i, j) \mid \varphi_i(j) \text{ is defined}\}$ is not recursive, since its characteristic function h is not effectively computable, by Theorem 5 of Section 4.3.

(iii) The set $\{(i, j) | \varphi_i = \varphi_j\}$ is not recursive, since its characteristic function *equivalent* is not effectively computable, by Theorem 7 of Section 4.3.

A number of basic results readily follow from the definition of a recursive set. A set is *cofinite* if its complement $\bar{A} = \mathbf{N} - A$ is finite.

4 Proposition. *If set A is finite or cofinite then A is recursive.*

PROOF. An always-convergent **while**-program can be written which stores a finite table listing the elements of A if A is finite, and the elements of $\bar{A}$ if A is cofinite. Note that "A is finite" includes the case where $A = \emptyset$, the empty set. □

5 Proposition. (i) *If A is a recursive set, then its complement $\bar{A}$ is also recursive.*

(ii) *If A_1 and A_2 are recursive sets, then so are their union $A_1 \cup A_2$ and intersection $A_1 \cap A_2$.*

PROOF. Left to the reader (Exercise 1). □

For the next and other results to come, we need to define new operations on sets. Let $R \subset \mathbf{N}^k$ be a k-ary relation, i.e., a set of k-tuples. The set

$$\{(a_1, \ldots, a_{i-1}, a_{i+1}, \ldots, a_k) | \exists a_i \in \mathbf{N} \text{ with } (a_1, \ldots, a_i, \ldots, a_k) \in R\}$$

is called the *projection of R along the ith coordinate*. If b is a fixed natural number, the set

$$\{(a_1, \ldots, a_{i-1}, a_{i+1}, \ldots, a_k) | (a_1, \ldots, a_{i-1}, b, a_{i+1}, \ldots, a_k) \in R\}$$

is called the *section of R at b along the ith coordinate*. (See Exercise 3 for a geometrical example of these concepts.) Finally, if $S \subset \mathbf{N}^l$ is an l-ary relation, then the (Cartesian) *product* of R and S, denoted by $R \times S$, is as usual the set

$$\{(a_1, \ldots, a_k, b_1, \ldots, b_l) | (a_1, \ldots, a_k) \in R \text{ and } (b_1, \ldots, b_l) \in S\}$$

6 Proposition. (i) *A section of a recursive relation is recursive.*

(ii) *The product $R \times S$ of two recursive relations is itself recursive.*

PROOF. Left to the reader (Exercise 4). □

On the other hand, *a projection of a recursive relation is not necessarily recursive.*

7 Example. Let $S \subset \mathbf{N}^3$ be the "step-counting" relation, where $(x, y, z) \in S$ $\Leftrightarrow \mathbf{P}_x$ on input y halts within z steps.

The characteristic function of S is the ψ_5 shown to be total computable in Section 4.1. Thus S is a recursive subset of $\mathbf{N}^3$. But the projection of S along the third coordinate equals

$$\{(x, y) \mid \varphi_x(y)\downarrow\} = \{(x, y) \mid \exists z \text{ such that } \mathbf{P}_x \text{ on input } y \text{ halts within } z \text{ steps}\}$$

and we know that this set is *not* recursive.

Recursively enumerable sets

A set is *recursively enumerable* if there is an algorithm for listing its members. We already encountered this idea in Section 3.1, where we explained our scheme for effectively enumerating all **while**-programs; and again in Section 4.3, when we established the nonexistence of an algorithm for listing the total computable functions. Although the members of a recursively enumerable set can be thought of as appearing in a definite order over time, different algorithms will give enumerations in different orders. Thus the order of enumeration is generally unimportant in determining if a set is recursively enumerable. We allow the possibility of repetitions in the listing of a recursively enumerable set.

To formalize the idea of an effective enumeration, we introduce the following definition.

8 Definition. A set $A \subset \mathbf{N}$ is *partially decidable* or *recursively enumerable* if either $A = \emptyset$ or $A = \text{RAN}(f)$, for some total computable function $f: \mathbf{N} \to \mathbf{N}$. When $A = \text{RAN}(f)$, the function f is called an *enumerating function* for the set A, since an effective enumeration of the members of A is simply $f(0), f(1), f(2), \ldots, f(i), \ldots, i \in \mathbf{N}$.

Here is the intuition for the term "partially decidable". Use f to provide an effective enumeration $f(0), f(1), f(2), \ldots$ for A. To decide if m belongs to A, we wait until we reach an n for which $f(n) = m$. However, this method is partial because it never lets us decide that an m does *not* belong to A.

As with recursive sets, we say a subset $A \subset \mathbf{N}^k$, $k > 1$, is recursively enumerable if $A = \text{RAN}(f)$ for some total computable function $f: \mathbf{N} \to \mathbf{N}^k$ (e.g., computed by a **while**-program for which we read a k-vector result from variables $X1, X2, \ldots, Xk$). For convenience we henceforth take recursively enumerable sets as subsets of $\mathbf{N}$, unless explicitly stated otherwise.

The expression "recursively enumerable" is conveniently abbreviated to r.e. And instead of "partially decidable" and "recursively enumerable", we often use the (perhaps more intuitive) terms "effectively enumerable" and "effectively listable".

9 Examples. (i) The most trivial example of an r.e. set is $\mathbf{N}$ itself. An enumerating function for $\mathbf{N}$ is simply the identity function $id:\mathbf{N}\to\mathbf{N}$, which is clearly computable.

(ii) The set of all prime natural numbers is r.e. We leave it to the reader to find an enumerating function for this set and a **while**-program that evaluates it.

(iii) Other examples of r.e. sets are the set of all perfect squares, the set of all Fibonacci numbers, and the set of all factorial numbers.

So far all our examples of r.e. sets are also recursive. The results to be given shortly, however, should make clear the following basic distinction: For a recursive set A, there is an effective procedure to determine whether any x is a member of A; whereas for a r.e. set B, there exists a *partial* procedure which gives the correct answer when $x \in B$, but may not give an answer if $x \notin B$. Recall the intuition behind the Halting Problem. We can always verify that $\varphi_x(x)\downarrow$ by running $\mathbf{P}_x$ on x. However, we cannot use this method to verify that $\varphi_x(x)\uparrow$, and indeed for a given x, a method of verification may not exist.

We first show that the recursiveness of a set implies its recursive enumerability.

10 Proposition. *Any recursive set is also recursively enumerable.*

PROOF. If $A = \emptyset$, then A is r.e. by definition. If A is nonempty, let χ_A be its characteristic function which is evaluated by an always-converging **while**-program, and let a_0 be a fixed element of A. An enumerating function for A is computed by the following program:

begin
  if $\chi_A(X1) \neq 1$ **then** $X1 := a_0$
end

This program computes the total function $f:\mathbf{N}\to\mathbf{N}$ with

$$f(n) = \begin{cases} n, & \text{if } n \in A; \\ a_0, & \text{if } n \notin A; \end{cases}$$

and clearly $\mathrm{RAN}(f) = A$. □

Before going further, we explicitly discuss the technique of *dovetailing*, which will play an important role in proving that certain sets are recursively enumerable. In the discussion to follow, **while**-programs will be used as devices to list (generally infinite) sequences of natural numbers. So far we have used them solely as "computers of functions". Now we also want to consider them as "enumerators of sets". Typically, if P is a k-variable **while**-program computing a total function $f:\mathbf{N}\to\mathbf{N}$, then we define a new

while-program Q:

```
begin n := 0;
    while true do
        begin X1 := n; X2 := 0; . . . ; Xk := 0;
            P;
            output X1;
            n := succ(n)
        end
end
```

In this program, n is a new variable not in $X1, \ldots, Xk$, and *true* is a short-hand notation for the test "$X1 = X1$?". The only instruction which is neither a basic one nor a macro according to the language of **while**-programs, is "*output* $X1$". We agree however that "*output* $X1$" is not formally part of program Q, and is only included as a metalinguistic reminder that Q is used as an enumerator. We must look at the value stored in $X1$ at the end of every iteration of the **while-do** loop, in order to know what set is being enumerated by Q.

With this convention, Q is some $\mathbf{P}_i$, i.e., program i in our standard enumeration of **while**-program. While the function φ_i computed by Q is empty, because it never halts, the set enumerated by Q is exactly $\{f(0), f(1), \ldots, f(n), \ldots\} = \mathrm{RAN}(f)$.

The preceding construction of Q as an enumerator requires that program P computes a total unary function φ_P. Note that if φ_P were not total, the program for Q would never exit the **while-do** loop for the first n for which $\varphi_P(n)$ is not defined. We now generalize this construction so that it can list (not necessarily in increasing order of n) all the defined values of $\varphi_P(n)$ even when φ_P is not total.

Suppose that we are given the following computations in an effectively enumerable list

$$C_0, C_1, C_2, \ldots, C_n, \ldots,$$

and we wish to generate the results of all these computations.[1] [$C_0, C_1, C_2, \ldots$ will be typically the computations of $\Phi(e,0), \Phi(e,1), \Phi(e,2), \ldots$ for some fixed e, or the computations of $\Phi(0,0), \Phi(1,1), \Phi(2,2), \ldots$.] An obvious approach would be simply to run each computation in turn:

```
begin n := 0
    while true do
        begin output result of C_n; n := succ(n) end
end
```

But this approach fails unless every computation C_n is terminating. For, suppose C_3 terminated, but C_2 did not. The above program would never

[1] Review our definition of a computation in Section 2.3.

leave C_2 to get on to return C_3's output. The solution to this quandary is to run each computation C_n for some finite number of steps, then go on to later computations, returning again and again if the given C_n has not terminated. In other words, we *dovetail* or interweave the computations of all the C_n's.

To do this, we need the idea of "step-counting" for an algorithm, as discussed in Example 5 of Section 4.1. First, recall from Section 3.4 the enumeration $\tau : \mathbf{N}^2 \to \mathbf{N}$ of all pairs of natural numbers with $\tau(x, y) = n$, with $x = \pi_1(n)$ and $y = \pi_2(n)$. Then our new algorithm for finding the output of all the C_n's is as follows:

```
begin n := 0;
    while true do
        begin x := π1(n);
            y := π2(n);
            if Cx halts within y steps then output result of Cx;
            n := succ(n)
        end
end
```

In this way, if C_x ever halts, it will do so in y_0 steps for some finite y_0, and so the above algorithm will output the result of C_x at stage $n = \tau(x, y)$ for every $y \geqslant y_0$.

In summary, *dovetailing* a (possibly finite but usually infinite) list of computations is a serial version of the strategy of carrying them out *in parallel* so as to avoid the possibility of being trapped in a nonterminating computation.

We now apply this technique to show that the converse of Proposition 10 is false, i.e., that there are r.e. sets which are not recursive.

11 Proposition. *The set* $K = \{i \mid \varphi_i(i)\downarrow\}$ *is* r.e. *but not recursive.*

Proof. The unsolvability of the Halting Problem tells us that the set K cannot be recursive. We next describe an algorithm for effectively listing the members of K. The algorithm proceeds by *dovetailing*, with C_i taking the form of computing $\varphi_i(i)$, i.e., the computation of $\mathbf{P}_i$ on input i.

```
begin n := 0;
    while true do
        begin x := π1(n); y := π2(n);
            if Px halts on x within y steps then output φx(x);
            n := succ(n)
        end
end
```

It is easy to see that the above algorithm will eventually list every member of K, since if $\mathbf{P}_i$ on input i halts in j steps, then i will be placed in K when n equals $\tau(i, j)$. If we now define the function f as

$$f(i) = \text{“the } (i+1)\text{st element listed by the above algorithm”},$$

for all $i \in \mathbf{N}$, f is total computable and such that $\mathrm{RAN}(f) = K$. □

We recall, from Section 1.3, the Cantor Diagonal Argument which established that the set of real numbers is not countable. We write $\aleph_0$ (aleph null) for the cardinality of the natural numbers, and we then write $2^{\aleph_0}$ for the cardinality of the set of subsets of $\mathbf{N}$. Now the characteristic function of $A \subset \mathbf{N}$ can be represented as binary real number $0 \cdot \chi_A(0)\chi_A(1) \cdots \chi_A(n) \cdots$. It is thus easy to see that Cantor's argument establishes that the set of subsets of $\mathbf{N}$ is uncountable. Yet there are only $\aleph_0$ r.e. subsets of $\mathbf{N}$, one (not necessarily distinct one) for every total computable function $f: \mathbf{N} \to \mathbf{N}$, in addition to the empty set $\emptyset$. Thus there must exist subsets of $\mathbf{N}$ which are not even r.e. We now exhibit a particular set which cannot be recursively enumerated.

12 Proposition. *The set* $\{i \mid \varphi_i : \mathbf{N} \to \mathbf{N} \text{ total}\}$ *is not* r.e. *and, a fortiori, not recursive.*

Proof. This is just Corollary 3 of Section 4.3, namely, that there is no effective enumeration of all the total computable functions. □

We can extend to the class of r.e. sets some of the closure properties established earlier for the class of recursive sets. For example, r.e. sets are closed under union and intersection (Exercise 5). But r.e. sets also have a property not readily shared by recursive sets: We can use the indexing of the computable functions as an indexing of the r.e. sets thereby obtaining a natural effective enumeration of all r.e. sets. We shall take up this important property in Section 7.2, but first we explore some of the elementary relationships between recursive and r.e. sets.

Recursive sets in terms of r.e. sets

Although we have shown that recursiveness implies recursive enumerability, but not the other way around, the next proposition establishes one way of characterizing the first notion in terms of the second.

13 Proposition. *A set $A \subset \mathbf{N}$ is recursive if and only if both A and its complement $\bar{A}$ are recursively enumerable.*

PROOF. If A is recursive, we have already noted in Proposition 5 that $\bar{A}$ is also recursive. And if A and $\bar{A}$ are recursive, then they are also r.e., by Proposition 10.

Now for the opposite implication: If A and $\bar{A}$ are r.e., we want to show that A is in fact recursive. If either A or $\bar{A}$ is the empty set, then A is immediately recursive.

If neither $A = \emptyset$ nor $\bar{A} = \emptyset$, then $A = \text{RAN}(f)$ and $\bar{A} = \text{RAN}(g)$, for some total computable functions f and g. An always-converging algorithm for deciding whether an arbitrary x is in A is as follows:

> "List in succession the elements $f(0), g(0), f(1), g(1), \ldots, f(i), g(i), \ldots,$ for all $i \in \mathbf{N}$; if x appears as a value of f, then $x \in A$, and if x appears as a value of g, then $x \notin A$".

The correctness of this algorithm follows from the fact that $\text{RAN}(f) \cap \text{RAN}(g) = \emptyset$ and $\text{RAN}(f) \cup \text{RAN}(g) = \mathbf{N}$. □

With the preceding result, it becomes an easy matter to prove the nonrecursive enumerability of certain sets. An example follows.

14 Corollary. *The set* $\bar{K} = \{i \mid \varphi_i(i)\uparrow\}$ *is not* r.e. *and, a fortiori, not recursive.*

PROOF. We already know from Proposition 11 that the set $K = \{i \mid \varphi_i(i)\downarrow\}$ is r.e. but not recursive. Its complement $\bar{K}$ therefore cannot be r.e., for otherwise K would then be recursive, by the proposition above. □

For the next result we need the following definition: The set $A \subset \mathbf{N}$ is said to be *recursively enumerable in increasing order* if it has a strictly increasing enumerating function f; that is, $i < j$ implies $f(i) < f(j)$.

15 Proposition. *An infinite set* $A \subset \mathbf{N}$ *is recursive if and only if* A *is* r.e. *in increasing order.*

PROOF. If the set A is infinite and recursive, it should be obvious that we can effectively enumerate it in increasing order: We start counting the natural numbers $0, 1, 2, \ldots, i, \ldots,$ according to their usual ordering and, as we count them, we also compute the value $\chi_A(0), \chi_A(1), \chi_A(2), \ldots, \chi_A(i), \ldots,$ where χ_A is the characteristic function of A. Whenever $\chi_A(i) = 1$, we include i in the listing of A. Since A is infinite, this informal procedure may be formalized to define a strictly increasing enumerating function for A.

Conversely, let f enumerate the set A in increasing order. To decide whether an arbitrary x is in A, we generate the range of f until a number $\geqslant x$ appears. This is guaranteed to happen since f is strictly increasing. Then $x \in A$ just in case x has already appeared in the listing of the range of f. We thus have an effective procedure for deciding membership in A, and A is recursive. □

According to the preceding result, the recursiveness of infinite sets is equivalent to their recursive enumerability in increasing order. This explains why the examples of r.e. sets given at the beginning of this section are all recursive. In each case—the set **N**, the set of primes, the set of perfect squares, the set of Fibonacci numbers, and the set of factorials—we can find a strictly increasing enumerating function.

We now give a straightforward application of the preceding proposition. Other applications are in the exercises.

16 Corollary. *Every infinite* r.e. *set has an infinite recursive subset.*

PROOF. Let $A \subset \mathbf{N}$ be the infinite r.e. set under consideration, and $f: \mathbf{N} \to \mathbf{N}$ an enumerating function for A. By the preceding proposition, it suffices to show that A has an r.e. subset B in increasing order. B is generated according to the following procedure.

Stage 1. Compute $f(0)$, and then place it in the list of B.

Stage 2. Compute $f(1)$. If $f(1) > f(0)$, place $f(1)$ in the list of B, otherwise discard $f(1)$.

⋮

Stage $(n+1)$. Compute $f(n)$. If $f(n) > max(f(0), \ldots, f(n-))$, place $f(n)$ in the list of B, otherwise discard $f(n)$.

⋮

The above described algorithm will go on generating new elements of B forever, because $A = \mathrm{RAN}(f)$ is infinite. B is clearly r.e. in increasing order. □

EXERCISES FOR SECTION 7.1

1. Show that if A_1 and A_2 are recursive, then so are $\overline{A}_1, A_1 \cup A_2, A_1 \cap A_2$.
2. Is the following set recursive:

$$\{\tau(i,x) \mid \text{the } i\text{th position in the decimal expansion of } \pi \text{ is the digit } x\}?$$

3. Sketch the relation $\{(x, y) \mid 5 \leqslant x \leqslant y\}$ in the plane. What is the projection of the relation along the coordinate x? What is the section at 100 along this coordinate?
4. Show that sections and products of recursive relations are themselves recursive.

5. Show that the r.e. sets are closed under union and intersection. (For infinite union and infinite intersection, see Exercise 11 of Section 7.2.)

6. (i) Show that every infinite r.e. set has a one-to-one enumerating function.
   (ii) Show that a nonempty set is recursive if and only if it has an enumerating function in nondecreasing order.
   (iii) Suppose A is an infinite r.e. set with enumerating function f, such that $f(2n+3) > f(2n+1)$ and $f(2n+2) > f(2n)$ for all $n \geqslant 0$. Show that A must be recursive.
   (iv) Suppose A is an infinite r.e. set with enumeration function f which satisfies the condition: There is a constant $c \in \mathbf{N}$ such that for all $i, j \geqslant c$, if $i < j$, then $f(i) < f(j)$; that is, f is strictly increasing over the set $\{c, c+1, c+2, \ldots\}$. Show that A is recursive.

7. (i) Show that every infinite recursive set has both a nonrecursive r.e. subset and a non-r.e. subset.
   (ii) Show that every infinite r.e. set has both a nonrecursive r.e. subset and a non-r.e. subset.
   (iii) Exhibit a non-r.e. set which has a recursive subset, a nonrecursive r.e. subset, and a non-r.e. subset. (*Hint*: Consider the set $\{i \mid \varphi_i \text{ is total}\}$.) Interestingly, as shown in Section 8.2, there are non-r.e. sets which do not have infinite recursive subsets, nor do they have nonrecursive r.e. subsets— i.e., all their infinite subsets are themselves non-r.e.

8. (Yasuhara) Is each of the following true or false? Prove carefully.

   (a) If f is a one-to-one total computable function such that RAN(f) is recursive, and if g is a total computable function such that for all $n \in \mathbf{N}$, $g(n) \leqslant f(n)$, then RAN(g) is a recursive set.
   (b) If f is a one-to-one total computable function such that RAN(f) is recursive, and if g is a total computable function such that for all $n \in \mathbf{N}$, $g(n) \geqslant f(n)$, then RAN(g) is a recursive set.

9. Explicitly using the dovetailing technique, show that each of the following sets is recursively enumerable:

   (a) $\{i \mid \varphi_i \text{ is not empty}\}$;
   (b) $\{i \mid \varphi_i \text{ is not one-to-one}\}$;
   (c) $\{i \mid \varphi_i \text{ is not a constant function}\}$;
   (d) $\{n \mid a \in \mathrm{DOM}(\varphi_n)\}$, for some fixed number a;
   (e) $\{n \mid a \in \mathrm{RAN}(\varphi_n)\}$, for some fixed number a.

10. Can we replace the dovetailing algorithm in the proof of Proposition 11 by the following algorithm?

```
begin n := 0;
    while true do
        begin x := excess(n);
          if P_x halts on x within n steps then output φ_x(x);
          n := succ(n)
        end
end
```

where *excess* is the excess-over-a-square function defined in Exercise 6 of Section 2.3. Justify your answer carefully.

11. (a) Exhibit a computable function $\theta : \mathbf{N} \to \mathbf{N}$ such that for any total computable function $f : \mathbf{N} \to \mathbf{N}$, there exists an integer n such that $\theta(n)\downarrow$ but $\theta(n) \neq f(n)$. Indicate briefly why your function works.
    (b) Given θ as in (a), and a *fixed* total computable function f, show that the set

$$A = \{ n \mid \theta(n)\downarrow \text{ and } f(n) \neq \theta(n) \}$$

is r.e. but not recursive (*Hint*: Define the function

$$\zeta(n) = \begin{cases} \theta(n), & \text{if } n \in A; \\ f(n), & \text{if } n \notin A. \end{cases}$$

Show that if A is recursive then ζ is a total computable function—which in turn contradicts the property satisfied by θ in (a).)

7.2 Indexing the Recursively Enumerable Sets

Having defined the recursive sets as the sets that have computable characteristic functions, we next argue that there is no direct way for effectively enumerating them. For, if $\{\chi_i : \mathbf{N} \to \{0,1\} \mid i \in \mathbf{N}\}$ were an effective listing of all the characteristic functions of the recursive subsets of $\mathbf{N}$, then the function $\hat{\chi}(i) = 1 \dot{-} \chi_i(i)$ would be a characteristic function not in the given enumeration and, by Church's Thesis, computable—which, by our standard diagonalization argument, is a contradiction. (See, however, Exercise 10 part (d) at the end of this section for a complementary discussion.)

By contrast, there is a natural effective enumeration of all r.e. sets, although this result is not immediately obvious. Indeed, since the r.e. sets are the ranges of total computable functions (plus the empty set $\emptyset$), and since the total computable functions are *not* recursively enumerable, the recursive enumerability of the r.e. sets may even seem unlikely.

Fortunately, there are alternative criteria for recursive enumerability, which readily provide us with several ways to enumerate the r.e. sets effectively. We now undertake to present some of these criteria and their consequences. However, there does not exist an alternative characterization that gives us a straightforward method for effectively enumerating recursive sets.

Alternative Characterizations of Recursive Enumerability

Using the dovetailing technique we now prove a fundamental result.

1 Theorem. (i) *A set A is recursively enumerable if and only if it is the domain of definition of a (in general, partial) computable function.*

(ii) *A set A is recursively enumerable if and only if A is the range of a (in general, partial) computable function.*

PROOF. (i) If A is the empty set, then it is the domain of the nowhere-defined computable function. If A is a nonempty r.e. set, let $A = \text{RAN}(f)$ for some total computable function f, and define the function $\theta : \mathbf{N} \to \mathbf{N}$ as

$$\theta(i) = \begin{cases} 1, & \text{if } i \in \text{RAN}(f); \\ \perp, & \text{otherwise.} \end{cases}$$

Clearly, $\text{DOM}(\theta) = \text{RAN}(f) = A$, and θ is computed by the **while**-program

begin $X2 := X1; X1 := 1; n := 0;$
  while $X2 \neq f(n)$ **do** $n := succ(n)$
end

where we use a macro for f that does not change the values of $X1$, $X2$, or n.

For the opposite implication of part (i), assume $A = \text{DOM}(\theta)$ for a computable function $\theta : \mathbf{N} \to \mathbf{N}$. If $\text{DOM}(\theta) = \emptyset$, then A is recursive and therefore r.e. as well. Otherwise fix some element a_0 of A and use the program

begin $x := \pi_1(X1);\ y := \pi_2(X1);$
  if *computation of* $\theta(x)$ *halts within* y *steps* **then** $X1 := x$
  else $X1 := a_0$
end

Clearly this computes a total function f for which $\text{RAN}(f) = \text{DOM}(\theta)$.

(ii) If A is the empty set, then A is the range of the nowhere-defined computable function. If A is nonempty r.e. set, $A = \text{RAN}(f)$ for some total computable function f, which is in particular a partial computable function.

For the opposite implication, let $A = \text{RAN}(\theta)$ for some partial computable function $\theta : \mathbf{N} \to \mathbf{N}$. The same technique used in part (i) will work here —except that instead of using input $\tau(x, y)$ to yield x as the value of f whenever $\theta(x)$ converges, we now use $\theta(x)$ itself as the result; still using a_0 otherwise. □

The above characterizations give us two different enumerations for the r.e. sets. Let $\varphi_0, \varphi_1, \ldots, \varphi_i, \ldots$ be any effective enumeration of the computable functions of one variable. Then

$$\text{DOM}(\varphi_0), \text{DOM}(\varphi_1), \ldots, \text{DOM}(\varphi_i), \ldots$$

is an effective enumeration of all the r.e. subsets of $\mathbf{N}$, as is

$$\text{RAN}(\varphi_0), \text{RAN}(\varphi_1), \ldots, \text{RAN}(\varphi_i), \ldots .$$

As shown in the next proposition, these two indexings of the r.e. sets are effectively intertranslatable.

2 Proposition. *There are total computable functions* $t_1 : \mathbf{N} \to \mathbf{N}$ *and* $t_2 : \mathbf{N} \to \mathbf{N}$ *such that for all* $i \in \mathbf{N}$:

(1) $\mathrm{DOM}(\varphi_i) = \mathrm{RAN}(\varphi_{t_1(i)})$,
(2) $\mathrm{RAN}(\varphi_i) = \mathrm{DOM}(\varphi_{t_2(i)})$.

PROOF. (1) We define the program-rewriting function t_1 by letting $t_1(i)$ be the index of the following **while**-program, which as usual uses variable $X1$ for both input and output:

begin $X2 := \Phi(i, X1)$ **end**

$\mathbf{P}_{t_1(i)}$ converges on some input a if and only if $\mathbf{P}_i$ converges on a. Furthermore, if $\mathbf{P}_{t_1(i)}$ halts on a, its output is a. Hence $\mathrm{DOM}(\varphi_i) = \mathrm{RAN}(\varphi_{t_1(i)})$.

(2) We define the program-rewriting function t_2 by letting $t_2(i)$ be the index of the **while**-program

```
begin n := 0; X2 := succ(X1);
    while X1 ≠ X2 do
        begin x := π1(n); y := π2(n);
          if Pi halts on x within y steps then X2 := Φ(i, x);
          n := succ(n)
        end
end
```

The lower-case variables in this program, n, x, and y, are assumed to be Xj's different from $X1$ and $X2$. Variable $X1$ is used for both input and output.

$\mathbf{P}_{t_2(i)}$ is a dovetailing algorithm which, on input a, searches for a value b (if any) such that $a = \varphi_i(b)$. Hence $\mathrm{RAN}(\varphi_i) = \mathrm{DOM}(\varphi_{t_2(i)})$. □

The translation functions t_1 and t_2 above are one-to-one (why?), but not necessarily onto. The question of whether they can be made onto is taken up in Exercise 6.

The common practice has been to take the effective enumeration of domains of computable functions, $\{\mathrm{DOM}(\varphi_i)\}$, as a basis for indexing the r.e. sets.

3 Definition. Let $\varphi_0, \varphi_1, \varphi_2, \ldots$ be our standard effective enumeration of the computable functions from $\mathbf{N}$ to $\mathbf{N}$. We say that $\mathrm{DOM}(\varphi_i)$ is the *ith recursively enumerable set* and that i is an *r.e. index* for $\mathrm{DOM}(\varphi_i)$. We denote $\mathrm{DOM}(\varphi_i)$ by W_i, so that

$$W_0, W_1, \ldots, W_i, \ldots$$

is our standard effective enumeration of the r.e. subsets of $\mathbf{N}$. In a similar way, for each $j > 1$, we can define $W_i^{(j)} = \mathrm{DOM}(\varphi_i^{(j)})$ to be the *ith recursively enumerable j-ary relation* on **N**, and denote our standard effective enumeration of the r.e. j-ary relations by $W_0^{(j)}, W_1^{(j)}, \ldots, W_i^{(j)}, \ldots$.

It is worth noting that the set W_i may be viewed as the set *accepted* by the ith program in our standard enumeration of **while**-programs computing functions from **N** to **N**: $\mathbf{P}_i$ *accepts* n just in case $\mathbf{P}_i$ halts on input n.

OPERATIONS ON r.e. SETS

There are several important consequences of the characterization of recursive enumerability given in Definition 3. Those to be presented next are closure properties of the class of r.e. sets; other properties are to be found in the exercises.

Given a set $A \subset \mathbf{N}$ and a partial function $\theta : \mathbf{N} \to \mathbf{N}$, we define the *image* of A *under* θ, denoted by $\theta(A)$, to be the set

$$\theta(A) = \{\theta(a) \mid a \in \mathrm{DOM}(\theta) \text{ and } a \in A\}.$$

If $A \subset \mathbf{N}$ and $\theta : \mathbf{N} \to \mathbf{N}$, we define the *preimage* of A *under* θ, denoted by $\theta^{-1}(A)$, to be the set

$$\theta^{-1}(A) = \{b \mid b \in \mathrm{DOM}(\theta) \text{ and } \theta(b) \in A\}.$$

Using the dovetailing technique it is easy to show that if θ is computable, then the image and preimage of an r.e. set under θ is r.e. (Exercise 2).

The next result shows how our machinery for indexing the r.e. sets allows us to establish more precise results about operations on r.e. sets.

4 Proposition. *There are total computable functions* $g_1 : \mathbf{N}^2 \to \mathbf{N}$ *and* $g_2 : \mathbf{N}^2 \to \mathbf{N}$ *such that, give* r.e. *set* W_j *and computable function* φ_i, *we have*:

(1) $\varphi_i(W_j) = W_{g_1(i,j)}$,
(2) $\varphi_i^{-1}(W_j) = W_{g_2(i,j)}$.

PROOF. (1) We first note that $\varphi_i(W_j) = \{\varphi_i(a) \mid a \in W_i \cap W_j\}$. We define the program-rewriting function $\hat{g} : \mathbf{N}^2 \to \mathbf{N}$ by letting $\hat{g}(i, j)$ be the index of the following **while**-program with input-output variable $X1$:

$$\mathbf{begin}\ X2 := \Phi(j, X1); X1 := \Phi(i, X1)\ \mathbf{end}$$

On input a, $\mathbf{P}_{\hat{g}(i,j)}$ first checks whether $\varphi_j(a)$ is defined, and if so, the value of $\varphi_j(a)$ is then stored in $X2$; then $\mathbf{P}_{\hat{g}(i,j)}$ checks whether $\varphi_i(a)$ is defined—and if it is, $\mathbf{P}_{\hat{g}(i,j)}$ returns $\varphi_i(a)$ in $X1$. Hence, $\varphi_i(W_j) = \mathrm{RAN}(\varphi_{\hat{g}(i,j)})$. To find an r.e. index for this set, we use the translation function t_2 of Proposition 2 to obtain

$$\varphi_i(W_j) = \mathrm{RAN}(\varphi_{\hat{g}(i,j)}) = \mathrm{DOM}(\varphi_{t_2(\hat{g}(i,j))}).$$

The desired function g_1 is defined by $g_1(i, j) = t_2(\hat{g}(i, j))$ for all i, j.

(2) A moment's thought will show that $\varphi_i^{-1}(W_j) = \mathrm{DOM}(\varphi_j \circ \varphi_i)$. Letting $g_2 : \mathbf{N}^2 \to \mathbf{N}$ be the total computable function satisfying $\varphi_{g_2(i,j)} = \varphi_j \circ \varphi_i$ (see

Example 2 of Section 4.2), we get $\varphi_i^{-1}(W_j) = \text{DOM}(\varphi_{g_2(i,j)}) = W_{g_2(i,j)}$, as required. □

Although the class of r.e. sets cannot be closed under complementation (why?), the union and intersection of r.e. sets are r.e. Using Church's Thesis, this result is easily established by a straightforward informal argument (Exercise 5 of Section 7.1). The next proposition establishes these results more formally, and in the process determines indices for $A \cup B$ and $A \cap B$ *uniformly* as a function of indices for A and B.

5 Proposition. *There are total computable functions join* : $\mathbf{N}^2 \to \mathbf{N}$ *and meet* : $\mathbf{N}^2 \to \mathbf{N}$ *such that, given* r.e. *subsets* W_i *and* W_j *of* $\mathbf{N}$,

$$W_i \cup W_j = W_{join(i,j)} \qquad \text{and} \qquad W_i \cap W_j = W_{meet(i,j)}.$$

Proof. The required function *meet* is none other than $\hat{g}$ in the proof of Proposition 4, since $W_i \cap W_j = \text{DOM}(\varphi_{\hat{g}(i,j)})$. The proof for the union requires us to find a program $\mathbf{P}_{join(i,j)}$ which computes

$$\varphi_{join(i,j)}(a) = \begin{cases} 1, & \text{if } \varphi_i(a) \text{ converges or } \varphi_j(a) \text{ converges;} \\ \perp, & \text{otherwise.} \end{cases}$$

where *join* is a total computable function. The details are left to the reader (Exercise 3). □

Recall that the class of recursive relations is closed under the operation of taking Cartesian products and that of taking sections. Similar results hold for the class of r.e. relations. But now, in contrast to Example 7 of Section 7.1, we observe that the projections of r.e. relations are also r.e.

6 Proposition

(i) *The product* $A \times B$ *of two* r.e. *relations* A *and* B *is itself* r.e.
(ii) *Any section of an* r.e. *relation is* r.e.
(iii) *Any projection of an* r.e. *relation is* r.e.

(*In each case, we can also compute an index for the resulting set from the indices of the given sets.*)

Proof. Exercise 4. □

The closure properties given in this section are extremely useful tools for proving that certain sets are r.e. We give one example below.

7 Example. Let $\varphi_0, \varphi_1, \varphi_2, \ldots$ be our standard enumeration of all the unary computable functions, and consider the set $A = \{i \mid 7 \in \text{RAN}(\varphi_i)\}$. To show that A is r.e., we first define an auxiliary relation $B \subset \mathbf{N}^3$.

$$B = \{(i, j, k) \mid \mathbf{P}_i \text{ on input } j \text{ halts within } k \text{ steps and } \varphi_i(j) = 7\}.$$

By Church's Thesis, B is recursive, and is therefore also r.e. It is easy to see that

$$\begin{aligned} A &= \{i \mid \text{there are } j \text{ and } k \text{ such that } \mathbf{P}_i \text{ on input } j \text{ halts within} \\ &\qquad k \text{ steps and } \varphi_i(j) = 7\} \\ &= \text{projection of } B \text{ along 2nd and 3rd coordinates.} \end{aligned}$$

By proposition 6, A is r.e.

Although the projection of a recursive set is not necessarily recursive, the projection of a recursively enumerable set is recursively enumerable, by Proposition 6. Hence, the projection of a recursive set is r.e. We now show that the converse is also true.

8 Theorem. *A relation is* r.e. *if and only if it is the projection of a recursive relation.*

PROOF. We already know that the projection of a recursive relation is r.e., by Proposition 6, part (iii).

For the converse, consider an arbitrary r.e. set $W_i = \text{DOM}(\varphi_i)$. (The proof for an arbitrary r.e. relation $W_i^{(j)} = \text{DOM}(\varphi_i^{(j)})$, for some $j > 1$, is similar.) We have

$$W_i = \{a \mid \text{there is a } k \text{ such that } \mathbf{P}_i \text{ halts on input } a \text{ within } k \text{ steps}\}.$$

The binary relation $R(a,k)$ defined by "$\mathbf{P}_i$ halts on input a within k steps" is recursive, by Church's Thesis (see also Example 5 of Section 4.1). Hence $a \in W_i$ if and only if $\exists k R(a,k)$, and W_i is the projection of a recursive relation. □

Theorem 8 is another useful characterization of recursive enumerability. A strengthened version shows that we can go uniformly *from* the r.e. index of a recursively enumerable relation A *to* the r.e. index of a recursive relation B such that A is the projection of B—and vice-versa (Exercise 7).

A j-ary relation $R \subset \mathbf{N}^j$ is said to be *Diophantine* if there is a Diophantine equation[2] $P(x_1, \ldots, x_j, y_1, \ldots, y_k)$ in $j + k$ variables such that:

$$(a_1, \ldots, a_j) \in R \text{ if and only if } \exists y_1 \cdots \exists y_k P(a_1, \ldots, a_j, y_1, \ldots, y_k) = 0,$$

for all $a_1, \ldots, a_j \in \mathbf{N}$ and where variables $y_1, \ldots, y_k$ range over $\mathbf{N}$. The proof of the next theorem, which involves some number theory, is beyond the scope of this book.

9 Theorem. *A relation is* r.e. *if and only if it is Diophantine.*

This theorem provides yet another important characterization of recursive enumerability. One direction (namely, "every Diophantine relation is

[2] The definition is given in Exercise 3 of Section 1.2.

r.e.") is relatively easy; the other direction ("every r.e. relation is Diophantine") is considerably more difficult, and was proved in 1970 by the Soviet mathematician Matijasevic. In fact Matijasevic's proof was the final breakthrough in a succession of results established in the 1950's and 1960's by the American mathematicians Davis, Putnam, and Robinson. Two interesting applications of Theorem 9 are given in Exercises 8 and 9.

Exercises for Section 7.2

1. Suppose A is recursive and f is a total computable function. Are $f(A)$ and $f^{-1}(A)$ recursive? r.e.? Justify your answer.

2. Suppose $A \subset \mathbf{N}$ is r.e. and $\theta : \mathbf{N} \to \mathbf{N}$ is a computable function. Show that $\theta(A)$ and $\theta^{-1}(A)$ are r.e. sets.

3. Prove Proposition 5.

4. Prove Proposition 6.

5. Show that there are r.e. indices m and n such that

   (i) $W_m = \{m, m^2, m^3, \ldots\}$,
   (ii) $W_n \subset \{n, n^2, n^3, \ldots\}$ *and* W_n is not recursive.

   (*Hint*: Use the Recursion Theorem.)

6. Consider the translation functions t_1 and t_2 of Proposition 2.

   (a) Show that t_1 and t_2 are not onto functions.
   (b) Prove the existence of translation funcitons $\hat{t}_1$ and $\hat{t}_2$ satisfying Proposition 2 which are both one-to-one and onto.

   (*Hint*: Review the proof of Theorem 4 of Section 6.2.)

7. Prove the following strengthened version of Theorem 8.

   (a) There is a total computable function $t : \mathbf{N} \to \mathbf{N}$ such that for all x, if $A = \{y \mid \exists z (y, z) \in W_x^{(2)}\}$, then $t(x)$ is a r.e. index for $A = W_{t(x)}$.
   (b) There is a total computable function $u : \mathbf{N} \to \mathbf{N}$ such that for all $x, u(x)$ is a r.e. index for a recursive binary relation $B = W_{u(x)}^{(2)}$ such that $W_x = \{y \mid \exists z (y, z) \in B\}$.

   (*Hint*: For part (b), define

$$\varphi_{u(x)}^{(2)}(y, z) = \begin{cases} \varphi_w^{(2)}(y, z), & \text{if } \varphi_w^{(2)}(y, z) = 1; \\ \bot, & \text{otherwise.} \end{cases}$$

   where $\varphi_w^{(2)}$ is the step-counting function of Example 5 of Section 4.1.)

8. (a) Show that every Diophantine relation is recursively enumerable. (*Hint*: Use Church's Thesis.)
   (b) Assuming that every r.e. relation is Diophantine (this is the "difficult" direction in Theorem 9), show that it is undecidable whether an arbitrary Diophantine equation has a solution in natural numbers. The definition of "Diophantine equation" is given in Exercise 3 of Section 1.2.

(c) Using part (b), show that the general Diophantine Problem (Hilbert's Tenth Problem), as defined in Exercise 3 of Section 1.2 is unsolvable. (*Hint*: For every Diophantine equation $P = 0$, show the existence of another Diophantine equation $P' = 0$ such that: $P = 0$ has a solution in natural numbers $\Leftrightarrow P' = 0$ has an integer solution. Use the well-known fact of number theory that every natural number is the sum of four squares.)

9. We wish to prove the following characterization of the r.e. sets: *A set A is* r.e. $\Leftrightarrow$ *A is the set of non-negative values taken by a polynomial* $p(x_1, \ldots, x_n)$ *with integer coefficients, as its variables* $x_1, \ldots, x_n$ *range over* **N**.

   (a) Prove the right-to-left direction of this theorem.
   (b) Prove the left-to-right direction of the theorem.

   (*Hint*: Use Theorem 9.)

10. (a) Show that if $A \subset \mathbf{N}$ is r.e. with r.e. index i, i.e., $A = W_i$, then A has in fact infinitely many r.e. indices.
    (b) Let $W_0, W_1, W_2, \ldots$ be our standard enumeration of the r.e. sets. Argue informally, but rigorously, that there is a total computable function $g: \mathbf{N} \to \mathbf{N}$ such that for every i:

    —the effective enumeration of the set $W_{g(i)}$ by dovetailing is in nondecreasing order;
    —if the effective enumeration of W_i by dovetailing is itself in nondecreasing order, then $W_{g(i)} = W_i$.

    (c) Show that for each i, $W_{g(i)}$ is recursive. (*Hint*: Consider separately the cases when $W_{g(i)}$ is empty or finite or infinite.)
    (d) Argue informally, but rigorously, that the class $\{W_{g(0)}, W_{g(1)}, \ldots, W_{g(i)}, \ldots\}$ is a r.e. class of all the recursive sets (it is a r.e. class in that it has a r.e. sequence of indices: $g(0), g(1), \ldots, g(i), \ldots$).

11. Let A be a r.e. set, and B a recursive set. Prove that while the set $\bigcup_{i \in A} W_i$ is necessarily r.e., the set $\bigcap_{i \in B} W_i$ need not be r.e. nor have r.e. complement. (*Hint*: First, find a total computable function g such that for all i, $W_{g(i)} = \{n \mid W_n$ has at least i members$\}$. Take B to be an infinite recursive subset of RAN(g), and show that $\bigcap_{i \in B} W_i = \bigcap_{i \in \mathbf{N}} W_{g(i)} = \{n \mid W_n$ is infinite$\}$.)

7.3 Gödel's Incompleteness Theorem

Historical perspective

It had been generally believed since the time of Euclid that the axioms of geometry are absolute truths whose validity is beyong dispute. As late as the eighteenth century, the German philosopher Kant proclaimed that the axioms of Euclidean geometry were given a priori to human intuition. This attitude received a severe jolt in the nineteenth century from the work of

Bolyai, Lobachewsky, and Riemann. These three mathematicians postulated systems of geometry that were *non-Euclidean* in that they denied the truth of the following *axiom* of the Euclidean scheme: "Given a line, and a point not on it, then there exists precisely one line through the given point parallel to the given line". In our own age, Euclidean absolutism has been completely discredited, and Einstein's theory of relativity has become dogma. In Einstein's view, the geometry of space is best described by a Riemannian non-Euclidean scheme in which parallel lines do not exist.

In other words, our present view not only contradicts Kant's view that the Euclidean axioms are given a priori to human intuition, but it even asserts that one of Euclid's axioms, the so-called parallel axiom, is actually untrue as a description of the universe. We now believe that Euclidean geometry is quite accurate enough to describe the spatial relations of our everyday lives, but not the spatial entirety of our universe. As a result, however, we are now faced with a genuine and important question which the Kantian viewpoint allowed us to dispose of effortlessly: How do we know that Euclidean geometry is consistent (i.e., free from contradictions)?

As if to underscore the depth and subtlety of this question, *Riemann showed that if Euclidean geometry was consistent, then his non-Euclidean geometry was consistent*. Here was a fine contretemps—not only had the consistency of the Euclidean scheme ceased to be a priori evident, but it was shown that its consistency implied the consistency of a rival scheme! One evident conclusion was that the study of axiomatic systems had to address the problem of consistency on a purely formal basis. A system's apparent accuracy in describing the "real world" was no longer sufficient, by itself, to establish the system's consistency.

The ramifications of this point of view rocked the mathematical world at the beginning of the twentieth century, when Cantor's newly developed theory of sets came under scrutiny. The set theory propounded by Cantor seemed completely consistent until Russell, among others, pointed out that this apparently "safe" system contained an annihilating paradox, which runs as follows: Consider the set M of all mathematicians. This set is not a mathematician, and so it does not belong to itself. However, the set of things talked about in this chapter is talked about in this chapter, so this set does belong to itself. Let us define N to be the set of all those sets which do not belong to themselves. Then

X belongs to N if and only if X does not belong to X.

So the set of mathematicians belongs to N, but the set of things described in this chapter does not. Does N belong to N? According to the above:

N belongs to N if and only if N does not belong to N.

A paradox! And so naive set theory is inconsistent. Now Russell avoided such contradictions by introducing his Theory of Types, but the point here is that consistency is a deep and elusive property, and not a trivially evident feature of a logical system.

Riemann had shown that his geometry is consistent if Euclidean geometry is consistent, and Hilbert showed that Euclidean geometry is consistent if arithmetic—the elementary theory of natural numbers—is consistent. Thus, in the early part of this century, mathematicians attempted to exhibit a consistent axiomatic system of arithmetic.

Now, a logical system consists of a collection of axioms, from which we obtain theorems by the repeated application of a number of rules of inference. The school of Formalists, led by Hilbert, decided that in their search for proofs of inconsistency in a logical system, they would ignore all questions of the truth and meaning of the axioms and theorems. Instead they regarded the axioms as mere strings of symbols and the rules of inference as purely syntactic replacement rules for obtaining new strings. They further decided to require that the rules of inference operate in a purely finitistic, effective manner, such as is exemplified very well in the operation of **while**-programs.

The Formalists were searching for a proof system which was *complete*, i.e., which was adequate for proving as theorems all true statements about the integers. Their program was wrecked by Gödel's Incompleteness Theorem, first expounded in his famous paper on formally undecidable theorems of Principia Mathematica and related systems [Gödel (1931)]. We now develop machinery which will let us appreciate and informally prove Gödel's result.

The incompleteness theorem

The well-formed formulas of first-order arithmetic, $\mathfrak{F}$, is the set of syntactically legal statements about $\mathbf{N}$ involving $+$, $*$, 0, *succ*, and $=$, as well as propositional connectives and the quantifiers $\forall$ and $\exists$. $\mathfrak{F}$ can be generated by a context-free grammer, and is therefore recursive (see Section 9.3). This language is termed *first-order* because $\forall$ and $\exists$ may range only over elements of $\mathbf{N}$ and not, as in a *second-order* arithmetic, over subsets of $\mathbf{N}$. The natural number n is identified with the nth successor of 0, which we abbreviate $\bar{n}$.

A formula with one free (i.e., unquantified) variable x may be true for some values of x and false for others. Thus

$$(x + x) * \bar{2} = 0$$

is true if $x = 0$, and false if $x > 0$. A wff $G \in \mathfrak{F}$ is a *sentence* if all its variables are quantified. A sentence is either true or false.

Now consider the set

$$\mathrm{TR} = \{ G \mid G \text{ is a sentence of } \mathfrak{F} \text{ which is true} \}.$$

TR includes

$$(\forall x)(x \neq 0 \textit{ implies } (\exists y)(\textit{succ}(y) = x))$$

but not

$$(\forall x)(\forall y)(\exists z)(x * y = \bar{2} * z).$$

In what sense is TR accessible to us by effective methods? The Formalists conjectured the existence of an effective procedure for deciding membership in the set; and Gödel's theorem settled their conjecture by proving it false.

To sketch (with hindsight) the Formalist proposal, we need to make more precise the notion of a formal system for deduction. A *proof system* $\mathcal{P}$ is a pair $\mathcal{P} = (\mathrm{AX}, \mathcal{R})$, where $\mathrm{AX} \subset \mathrm{TR}$ is a set of *axioms*, and $\mathcal{R}$ is a finite set of *rules of inference*. Some of the axioms (for most formulations) of proof systems for arithmetic are familiar to us from elementary geometry, e.g.,

$$(\forall x)(\forall y)(\forall z)((x = y \text{ and } y = z) \text{ implies } x = z);$$

other are specific to arithmetic:

$$(\forall x)(succ(x) \neq 0),$$

(which says that 0 is the successor of no natural number).

Rules of inference often include *modus ponens*: "From G and (G *implies* H) infer H," and also rules for quantifiers, such as "From $(\forall x)F(x)$ infer $F(c)$, where c is a constant".

More formally, a rule of inference $R \in \mathcal{R}$ is a $(k+1)$-ary relation on $\mathcal{F}$. If $(F_1, F_2, \ldots, F_{k+1}) \in R$, we say "from $F_1, \ldots, F_k$ infer F_{k+1}" or "F_{k+1} is *derivable* from $F_1, \ldots, F_k$". We also require that every rule of inference be *sound*: An inference rule R is sound if $(F_1, \ldots, F_k, F_{k+1}) \in R$, and $F_1, \ldots, F_k \in \mathrm{TR}$, implies $F_{k+1} \in \mathrm{TR}$ (sound rules preserve truth).

1 Definition. A sentence G is a *theorem* of $\mathcal{P} = (\mathrm{AX}, \mathcal{R})$ if G is an axiom, or G is derivable by some proof rule of $\mathcal{R}$ from previously proven theorems or from axioms. We write $\vdash_{\mathcal{P}} G$ to indicate that G is a theorem of $\mathcal{P}$, and we define

$$\mathrm{THM}(\mathcal{P}) = \{ G \mid G \text{ is a sentence in } \mathcal{F} \text{ and } \vdash_{\mathcal{P}} G \}.$$

Recapitulating, the Formalists attempted to construct a proof system $\mathcal{P}$ for which $\mathrm{THM}(\mathcal{P}) = \mathrm{TR}$. Such a system is called *complete* because every true sentence about arithmetic is provable. Since all rules of inference are sound, no false sentence is provable. Hence a proof system is *incomplete* if for some sentence G, neither it nor its negation is provable.

Notice that if we take $\mathrm{AX} = \mathrm{TR}$, then $\mathcal{P} = (\mathrm{TR}, \mathcal{R})$ is complete even if the set $\mathcal{R}$ of inference rules is empty. But this is no help, since we don't know which sentences are in TR to begin with. Accordingly we make the following definition, assuming that a standard Gödel numbering of $\mathcal{F}$ has been given.

2 Definition. Proof system $\mathcal{P} = (\text{AX}, \mathcal{R})$ is *recursive* if AX is a recursive subset of $\mathcal{F}$ and each R in $\mathcal{R}$ is a recursive subset of $\mathcal{F}^{k+1}$ for suitable k. We can now state Gödel's theorem.

3 Theorem (The Incompleteness Theorem). *Every recursive proof system* $\mathcal{P}$ *is incomplete. That is, if* $\mathcal{P}$ *is a recursive proof system, then* THM($\mathcal{P}$) *is a proper subset of* TR.

To outline the proof of the Incompleteness Theorem we need the following result.

4 Proposition. *If* $\mathcal{P}$ *is a recursive proof system, then* THM($\mathcal{P}$) *is* r.e.

PROOF. We begin by dovetailing the listing of the axiom set, AX, of $\mathcal{P}$. In the process, we interleave applications of the inference rules of $\mathcal{P}$ to this list, and place the newly proven theorems at the bottom of the list. We continue dovetailing in this manner, listing new axioms, deducing and then listing new theorems, making sure that we suitably alternate the two processes. □

Call a number-theoretic relation $A(x_1, \ldots, x_j)$ *definable* if there is a formula $F_A \in \mathcal{F}$ such that

$$A(k_1, \ldots, k_j) \text{ holds} \Leftrightarrow F_A(\bar{k}_1, \ldots, \bar{k}_j) \text{ is true.}$$

For example, the relation $x < y$ is definable by the formula $(\exists z)(z \neq 0$ *and* $x + z = y)$. Our next theorem is the one key technical result which we leave unproven.

5 Theorem. *The relation* $T(e, x, y)$, *which asserts "***while***-program e on input x halts in exactly y steps" is definable, say by formula* $B(e, x, y)$ *in* $\mathcal{F}$.

6 Corollary. *The relation* $x \in W_e$ *is definable.*

PROOF. The formula of $\mathcal{F}$ defining the relation "$x \in W_e$" is just $W(e, x) = (\exists y)B(e, x, y)$. □

With Corollary 6 in hand, we can now prove the Incompleteness Theorem.

PROOF OF THEOREM 3. Suppose some recursive proof system $\mathcal{P}$ is complete. Then every sentence $W(\bar{n}, \bar{m})$ is decided by $\mathcal{P}$: either $\vdash_{\mathcal{P}} W(\bar{n}, \bar{m})$ or $\vdash_{\mathcal{P}} \neg W(\bar{n}, \bar{m})$. Since THM($\mathcal{P}$) is r.e., we can therefore decide the General Halting Problem as follows. To tell if $n \in W_m$, simply list THM($\mathcal{P}$) until either $W(\bar{n}, \bar{m})$ or $\neg W(\bar{n}, \bar{m})$ turns up. By completeness, one sentence or the other must appear in the list of theorems. By the soundness of the inference rules, the sentence which turns up must be true. But this is a contradiction, since the General Halting Problem is unsolvable. □

Thus Gödel's result shows that no effective system of deduction can capture the set of true statements of arithmetic. A related result of comparable power was proved independently by Tarski in the early 1930's [Tarski (1932)]. Tarski showed that the set TR is not even definable in arithmetic.

CHAPTER 8

Computable Properties of Sets (Part 2)

Here we continue the previous chapter's discussion of computable properties of sets.

In Section 8.1 we prove Rice's Theorem and several of its important extensions. Rice's Theorem is a powerful tool for proving that certain decision problems are algorithmically unsolvable. Indeed most of the undecidability results presented so far are easy consequences of this result. In the context of actual computer programming, Rice's Theorem demonstrates that there is no uniform effective way for testing input-output behaviors of arbitrary programs. Of course, we have already seen in Chapter 5 that there do exist *restricted* classes of programs for which many properties are indeed effectively testable.

We have seen that certain sets are recursive, others are recursively enumerable without being recursive, and still others are not even recursively enumerable. This classification of sets can be refined further. For example, recursive sets can be classified according to the complexity (i.e., time and space requirements) of their decision procedures. The study of such "subrecursive" classification systems is an important aspect of theoretical computer science beyond the scope of this book. [The interested reader may consult Chapters 5 and 6 of Machtey and Young (1978).]

We make precise the notion of "degree of unsolvability" in Section 8.2, where we introduce several schemes for classifying nonrecursive sets.

8.1 Rice's Theorem and Related Results

To motivate Rice's Theorem, let us consider the set of indices $A = \{i \mid 7 \in \mathrm{RAN}(\varphi_i)\}$. Suppose $i \in A$, and suppose $\varphi_i = \varphi_j$. Then it follows that $j \in A$ as well. In other words, the index set A has the following property: Either all of the indices for a given computable function fall in A, or all fall in $\bar{A} = \mathbf{N} - A$. Rice's Theorem guarantees that any nontrivial partition of $\mathbf{N}$ with this property must be nonrecursive.

1 Definition. A set of indices $I \subset \mathbf{N}$ *respects functions* if $i \in I$ and $\varphi_i = \varphi_j$ imply $j \in I$. Put differently, if I contains program i, then I contains every other program j that computes the same function as program i.

It should be clear that if I respects functions, then so does its complement $\bar{I}$. Notice that the sets of indices $\{i \mid 7 \in \mathrm{RAN}(\varphi_i)\}$, $\{i \mid \varphi_i \text{ total}\}$, and $\{i \mid \varphi_i \text{ empty}\}$ all respect functions, while $K = \{i \mid \varphi_i(i)\downarrow\}$ does not. Indeed, by Example 5 of Section 6.1, there is an index e such that $W_e = \{e\}$, placing e in K, but for any $e' \neq e$ such that $\varphi_{e'} = \varphi_e$ we have $e' \notin K$.

2 Theorem (Rice's Theorem). *Suppose the index set $I \subset \mathbf{N}$ respects functions. Then I is recursive if and only if $I = \emptyset$ or $I = \mathbf{N}$.*

Proof. If $I = \emptyset$ or $I = \mathbf{N}$, then I is certainly a recursive set. For the converse, we assume that $\emptyset \neq I \neq \mathbf{N}$ and show that I cannot be recursive. Assume the contrary. Without loss of generality we may also assume that I contains all the indices for some computable function θ not equal to the empty (nowhere-defined) function ν, while $\bar{I}$ contains all the indices for ν. (If I does not satisfy this assumption then its complement $\bar{I}$ does.)

Define $\mathbf{P}_{f(i)}$ to be the program

$$\textbf{begin } X2 := \Phi(i,i); X1 := \theta(X1) \textbf{ end}$$

It should be clear that

$$\varphi_{f(i)} = \begin{cases} \theta, & \text{if } \varphi_i(i)\downarrow; \\ \nu, & \text{otherwise.} \end{cases}$$

Thus $f(i) \in I$ if and only if $\varphi_i(i)$ converges for all $i \in \mathbf{N}$. Hence if χ_I is the characteristic function of the index set I,

$$\chi_I(f(i)) = \begin{cases} 1, & \text{if } \varphi_i(i)\downarrow; \\ 0, & \text{otherwise.} \end{cases}$$

for all $i \in \mathbf{N}$. The function $\chi_I \circ f$ is thus none other than the function *halt* defined in Sections 1.2 and 3.1. Since it was proved that *halt* cannot be algorithmic, χ_I cannot be either. But this just says that I is not a recursive set. □

An index set $I \subset \mathbf{N}$ which respects functions defines a property of computable functions, and not just a property of programs, because it partitions the computable functions (and not just their indices) into two disjoint classes. Let us call a property of the computable functions an *input-output behavior* of the **while**-programs. Every index set which respects functions induces an input-output behavior, and vice versa.

Let us call an input-output behavior *nontrivial* if the corresponding index set is neither $\emptyset$ nor $\mathbf{N}$. For example, the property of producing an output for every possible input is a nontrivial input-output behavior, since some computable functions are total while others are not.

We can therefore restate Rice's Theorem as follows:

The set of all **while**-*programs satisfying a given nontrivial input-output behavior cannot be a recursive set.*

This is indeed a sweeping conclusion! It invalidates any attempt at finding an effective test for any nontrivial input-output behavior of arbitrary programs. However, notice that this principle only applies when the class of *all* possible programs is considered. For, as we have seen in Chapter 5, there is a program methodology that gives us techniques for effectively establishing formal properties of programs in specific instances.

Let us now illustrate Rice's Theorem with a few examples.

3 Example. By Rice's Theorem, each of the following sets of indices and their complements are undecidable (some already known to be so):

(1) $A_1 = \{i \mid \varphi_i \text{ is total}\}$;
(2) $A_2 = \{i \mid \varphi_i = f\}$, where f is a fixed total computable function;
(3) $A_3 = \{i \mid \varphi_i = \zeta\}$, where ζ is a fixed nontotal computable function;
(4) $A_4 = \{i \mid a \in \mathrm{DOM}(\varphi_i)\}$, for some fixed number a;
(5) $A_5 = \{i \mid \mathrm{DOM}(\varphi_i) = \emptyset\}$;
(6) $A_6 = \{i \mid \mathrm{DOM}(\varphi_i) \text{ is finite}\}$;
(7) $A_7 = \{i \mid a \in \mathrm{RAN}(\varphi_i)\}$, for some fixed number a;
(8) $A_8 = \{i \mid \mathrm{RAN}(\varphi_i) \text{ is finite}\}$;
(9) $A_9 = \{i \mid \mathrm{RAN}(\varphi_i) = \mathbf{N}\}$;
(10) $A_{10} = \{i \mid \varphi_i \text{ is one-to-one}\}$;
(11) $A_{11} = \{i \mid \varphi_i \text{ is a bijection}\}$.

For part (11), a bijection is a one-to-one, onto, total function.

Although Rice's Theorem immediately implies the undecidability of a wide class of decision problems, it does not establish, at least directly, the undecidability of the set $\{(i, j) \mid \varphi_i(j)\downarrow\}$ which is not a set of indices; nor does it directly establish the undecidability of K, which as an index set does not respect functions.

We next present a generalized version of Theorem 2 which will allow us to talk about the input-output behavior of pairs, triples, and, in general, n-tuples of programs.

4 Definition. A set $J \subset \mathbf{N}^n$ *respects functions* if $(a_1, \ldots, a_n) \in J$ and $\varphi_{a_1} = \varphi_{b_1}, \varphi_{a_2} = \varphi_{b_2}, \ldots, \varphi_{a_n} = \varphi_{b_n}$ imply $(b_1, \ldots, b_n) \in J$.

5 Theorem. *Suppose $J \subset \mathbf{N}^n$ respects functions. Then J is recursive if and only if $J = \emptyset$ or $J = \mathbf{N}^n$.*

PROOF. Exercise 6. □

6 Example. The following are not recursive:

(i) $A_{12} = \{(i, j) \mid \varphi_i = \varphi_j\}$;
(ii) $A_{13} = \{(i, j) \mid$ for every $a \in \mathrm{DOM}(\varphi_i) \cap \mathrm{DOM}(\varphi_j), \varphi_i(a) < \varphi_j(a)\}$;
(iii) $A_{14} = \{(i, j, k) \mid \varphi_i = \varphi_j \circ \varphi_k\}$.

Observe the different ways in which we have now established the undecidability of A_{12} (the unsolvability of the Equivalence Problem).

While we have shown that a number of input-output behaviors of **while**-programs are not effectively testable, we would like to establish the stronger result that some of these sets are not even partially decidable. Under certain circumstances we can do this using the following extension of Rice's Theorem.

7 Theorem. *Suppose the index set $I \subset \mathbf{N}$ respects functions, and furthermore suppose that there exists a computable function θ such that:*

(1) *$\{i \mid \varphi_i = \theta\} \subset I$, and*
(2) *there is a computable function $\hat{\theta}$ such that $\theta \leqslant \hat{\theta}$ and $\{i \mid \varphi_i = \hat{\theta}\} \subset \bar{I}$.*

Then the set I is not recursively enumerable.

Note that condition (2) forces $\hat{\theta} \neq \theta$, and therefore θ cannot be a total function.

PROOF. This is an amended version of the less trivial direction in the proof of Rice's Theorem. From Example 3 of Section 4.1, we know that the function $\zeta : \mathbf{N}^2 \to \mathbf{N}$ defined as

$$\zeta(i, j) = \begin{cases} \theta(j), & \text{if } \varphi_i(i) \text{ diverges;} \\ \hat{\theta}(j), & \text{if } \varphi_i(i) \text{ converges;} \end{cases}$$

is computable. Thus $\zeta(i, j) = \varphi_e(i, j)$ for some e, and by the s-m-n Theorem, we can infer the existence of a total computable f such tht $\zeta(i, j) = \varphi_{f(i)}(j)$. That is, $f(i) \in I$ if and only if $\varphi_i(i)$ diverges. Since $K = \{i \mid \varphi_i(i)$ converges$\}$, this can be rewritten as $\bar{K} = f^{-1}(I)$, but then if I were r.e., $\bar{K}$

would also be r.e. (see Proposition 4 of Section 7.2). We are forced to conclude that I cannot be r.e. . □

Consider again the set of all **while**-programs used to evaluate all computable functions from **N** to **N**. Let us define a *nontotal input-output behavior* of the **while**-programs to be any property that holds for some nontotal computable function θ but not for one of its computable extensions $\hat{\theta}$. A moment of reflection will show that Theorem 7 implies the following:

The set of all **while**-*programs satisfying a given nontotal input-output behavior cannot be recursively enumerated.*

This means that there are input-output behaviors which are not even partially testable. Notice that $A_1 = \{n \mid \varphi_n \text{ total}\}$ does not satisfy the hypotheses of Theorem 7, even though it too is not r.e. .

8 Example. Using Theorem 7, it is now an easy exercise to prove that the following sets of programs are not r.e.:

(1) $\bar{A}_1 = \{i \mid \varphi_i \text{ is not total}\}$;
(2) $\bar{A}_2 = \{i \mid \varphi_i \neq f\}$, where f is a total computable function;
(3) $A_3 = \{i \mid \varphi_i = \zeta\}$, where ζ is a nontotal computable function;
$\bar{A}_3 = \{i \mid \varphi_i \neq \zeta\}$, where ζ is a nontotal computable function such that $\mathrm{DOM}(\zeta) \neq \emptyset$;
(4) $\bar{A}_4 = \{i \mid a \notin \mathrm{DOM}(\varphi_i)\}$, for some fixed number a;
(5) $A_5 = \{i \mid \mathrm{DOM}(\varphi_i) = \emptyset\}$;
(6) $A_6 = \{i \mid \mathrm{DOM}(\varphi_i) \text{ is finite}\}$; → NOT r.e.
(7) $\bar{A}_7 = \{i \mid a \notin \mathrm{RAN}(\varphi_i)\}$, for some fixed number a;
(8) $A_8 = \{i \mid \mathrm{RAN}(\varphi_i) \text{ is finite}\}$;
(9) $\bar{A}_9 = \{i \mid \mathrm{RAN}(\varphi_i) \neq \mathbf{N}\}$;
(10) $A_{10} = \{i \mid \varphi_i \text{ is one-to-one}\}$;
(11) $\bar{A}_{11} = \{i \mid \varphi_i \text{ is not a bijection}\}$.

Another useful variation on Rice's Theorem follows.

9 Theorem. *Suppose the index set $I \subset \mathbf{N}$ respects functions, and furthermore suppose that there exists a computable function θ such that*

(1) $\{i \mid \varphi_i = \theta\} \subset I$, *and*
(2) $\{i \mid \varphi_i \leqslant \theta$ *and* $\mathrm{DOM}(\varphi_i)$ *finite*$\} \subset \bar{I}$ (*i.e., every program that computes a finite portion of θ is not in I*).

Then the set I is not recursively enumerable.

PROOF. Define the function $\zeta : \mathbf{N}^2 \to \mathbf{N}$ by

$$\zeta(i, j) = \begin{cases} \theta(j), & \text{if } \mathbf{P}_i \text{ does not halt on } i \text{ within } j \text{ steps;} \\ \perp, & \text{if } \mathbf{P}_i \text{ halts on } i \text{ within } j \text{ steps.} \end{cases}$$

By Church's Thesis, ζ is computable, so that $\zeta = \varphi_e$ for some e. By the s-m-n Theorem we can infer the existence of a total computable f such that $\zeta(i, j) = \varphi_{f(i)}(j)$. We now have

$$\begin{aligned} i \in K &\Leftrightarrow \mathbf{P}_i \text{ halts on } i \\ &\Leftrightarrow \mathbf{P}_i \text{ halts on } i \text{ in exactly } j \text{ steps, for some } j \\ &\Leftrightarrow \varphi_{f(i)}(x) = \theta(x) \text{ for } x < j \text{ and } \varphi_{f(i)}(x) = \bot \text{ for } x \geqslant j, \text{ for some } j \\ &\Leftrightarrow \varphi_{f(i)} \leqslant \theta \text{ and DOM}(\varphi_{f(i)}) \text{ finite} \\ &\Rightarrow f(i) \in \bar{I}. \end{aligned}$$

On the other hand, we also have

$$i \in \bar{K} \Leftrightarrow \mathbf{P}_i \text{ does not halt on } i \Leftrightarrow \varphi_{f(i)} = \theta \Rightarrow f(i) \in I.$$

(It is worth observing that the implication $[\varphi_{f(i)} = \theta \Rightarrow \mathbf{P}_i$ does not halt on $i]$ is generally true only because DOM(θ) is infinite.) We can now conclude that $i \in \bar{K}$ if and only if $f(i) \in I$ which, as in the proof of Theorem 7, means that I cannot be r.e. □

Let us define an *infinite input-output behavior* of the **while**-programs to be a property associated with some computable function θ but not with any finite function $\theta' \leqslant \theta$. (A function θ' is finite if its domain is finite; any finite function is computable—why?) Theorem 9 implies:

The set of all **while**-*programs satisfying a given infinite input-output behavior cannot be recursively enumerated.*

It is thus not even partially decidable whether an arbitrary **while**-program computes a given infinite function θ!

10 Example. Using Theorem 9, we can now prove that some of the sets mentioned in Example 3 or their complements are not r.e. (We mention here only the sets whose nonrecursive enumerability cannot be directly deduced from Theorem 7.)

(1) $A_1 = \{i \mid \varphi_i \text{ is total}\}$;
(2) $A_2 = \{i \mid \varphi_i = f\}$, where f is a computable total function;
(6) $\bar{A}_6 = \{i \mid \text{DOM}(\varphi_i) \text{ is infinite}\}$;
(8) $\bar{A}_8 = \{i \mid \text{RAN}(\varphi_i) \text{ is infinite}\}$;
(9) $A_9 = \{i \mid \text{RAN}(\varphi_i) = \mathbf{N}\}$;
(11) $A_{11} = \{i \mid \varphi_i \text{ is a bijection}\}$.

The cases which are not covered by either Theorem 7 or Theorem 9, namely, A_4, $\bar{A}_5$, A_7, and $\bar{A}_{10}$, are all recursively enumerable (Exercise 9 of Section 7.1).

Exercises for Section 8.1

1. Prove Rice's Theorem using the Recursion Theorem. (*Hint*: Let $\emptyset \neq I \neq \mathbf{N}$ be a set of indices which respects functions, with $i \in I$ and $j \notin I$, and define the function

$$f(x) = \begin{cases} i, & \text{if } x \notin I; \\ j, & \text{if } x \in I.) \end{cases}$$

2. A binary relation $R \subset \mathbf{N} \times \mathbf{N}$ is *single-valued* if $(a,b) \in R$ and $(a,c) \in R$ imply $b = c$. Show that the set $\{n \mid W_n^{(2)}$ is single-valued$\}$ is not r.e. in two different ways: (i) by showing that if it were r.e. then a computable characteristic function for K could be obtained, and (ii) by applying the results of this section.

3. (Rogers) Show that there is a total computable function $f: \mathbf{N} \to \mathbf{N}$ such that the set of fixed points of f (in the sense of the Recursion Theorem) is not r.e. (*Hint*: Define a program-rewriting function f such that the set of programs $\{i \mid \varphi_i = \varphi_{f(i)}\}$ respects functions and, in addition, satisfies the hypotheses of Theorem 7 or Theorem 9.)

4. Call a partial function $\zeta: \mathbf{N} \to \mathbf{N}$ *finite* if DOM(ζ) is finite. Consider the following set of programs:

$$I = \{i \mid \zeta \leqslant \varphi_i\}$$

where ζ is a finite function. Does I respect functions? Is I recursive? Is I r.e.? Is its complement $\bar{I}$ r.e.? Justify your answers.

5. A **while**-program computation cycles if at two points in the computation the program's state vectors are identical, following execution of the same instruction.[1] It follows that this state-vector configuration will be reached infinitely often. Is the set

$$\{n \mid \mathbf{P}_n \text{ with all variables initialized to 0 eventually cycles}\}$$

r.e.? Is this set recursive? Does Rice's Theorem apply in either case? Explain.

6. Prove Theorem 5.

7. Prove that $\{x \mid \varphi_x(x^2)\downarrow\}$ is not recursive. Does Rice's Theorem apply? Is this set r.e.?

8. Consider the following sets of indices:

   (a) $\{i \mid W_i$ cofinite$\}$,
   (b) $\{i \mid W_i$ recursive$\}$,
   (c) $\{i \mid W_i$ recursive, but neither finite nor cofinite$\}$.

   Does the set in each case respect functions? Is it recursive, r.e., or non-r.e.? Justify your answers.

9. Let θ be a unary computable function. Prove that the set of programs $I = \{i \mid \theta \leqslant \varphi_i\}$ is recursively enumerable if and only if θ is a finite function, i.e.,

[1] Review the definitions in Section 2.3.

DOM(θ) is finite. (*Hint*: For the left-to-right direction, use Theorem 7 or Theorem 9; for the right-to-left direction, use a dovetailing technique.)

10. We take a finite function ζ to be a single-valued relation of the form $\{(a_1, b_1), \ldots, (a_n, b_n)\}$ (see Exercise 2). Let us call a **while**-program P a *canonical program* for ζ if P is:

begin $X2 := X1$;
  if $X2 = a_1$ **then** $X1 := b_1$;
  if $X2 = a_2$ **then** $X1 := b_2$;
  ⋮
  if $X2 = a_n$ **then** $X1 := b_n$;
  if $X2 \neq a_1 \wedge \cdots \wedge X2 \neq a_n$ **then while** *true* **do begin end**
end

With P using variable $X1$ for both input and output, we clearly have $\zeta = \varphi_P$. Assuming that $a_1 < a_2 < \cdots < a_n$, P is the unique canonical program for ζ. If e is the index of **while**-program P, i.e., if $P = \mathbf{P}_e$, we shall also say that e is a *canonical index* for ζ. A program (or an index) is canonical only if it is a program (or an index) for a finite function.

(a) Argue that the set of all canonical programs (for unary finite functions) is r.e. Hence if I is the set of all canonical indices, then the set of functions $\{\varphi_i \mid e \in I\}$ is a r.e. class of all finite functions. Does I respect functions?
(b) Let program e be a fixed canonical program. Show that the set of programs $\{i \mid \varphi_i = \varphi_e\}$ is not recursively enumerable.
(c) Prove there is *no* total computable function $f: \mathbf{N} \to \mathbf{N}$ such that if program i computes a finite function, then $f(i)$ is (the index of) the canonical program for φ_i. Put differently, we cannot "uniformly go from indices in general to canonical indices". (*Hint*: The existence of such an f would contradict the fact that $\{i \mid \text{DOM}(\varphi_i) \text{ finite}\}$ is not r.e.)

11. We wish to prove a generalization of the result in Exercise 9, based on the notions introduced in the preceding exercise. Let $I \subset \mathbf{N}$ be a set of indices which respects unary functions. Then

I is recursively enumerable if and only if there is a r.e. *set A of canonical indices such that $I = \{i \mid \varphi_e \leqslant \varphi_i$ for some $e \in A\}$.*

(a) Show that if there is an r.e. set A of canonical indices such that $I = \{i \mid \varphi_e \leqslant \varphi_i$ for some $e \in A\}$, then I is recursively enumerable. (*Hint*: The required dovetailing argument is slightly more complicated than those discussed in the text.)
(b) Show that if I is recursively enumerable, then there is an r.e. set A of canonical indices such that $I = \{i \mid \varphi_e \leqslant \varphi_i$ for some $e \in A\}$. (*Hint*: Let $I = \text{DOM}(\varphi_n)$ and $f: \mathbf{N} \to \mathbf{N}$ be an enumerating function for the set of all canonical indices. Define $A = \text{DOM}(\varphi_n \circ f)$. Use Theorems 7 and 9 to show that $i \in I \Leftrightarrow \varphi_e \leqslant \varphi_i$ for some $e \in A$.)

8.2 A Classification of Sets

In this section we present a classification of nonrecursive sets, and show that some are more unsolvable than others in a sense to be made precise. We shall only scratch the surface of an extensive theory of "degrees of unsolvability", which is a natural follow-up of our study of elementary computability. For a wealth of further results see Rogers (1967).

PRODUCTIVE AND CREATIVE SETS

Most of our earlier proofs showing that certain sets are not recursive involved a "reduction" from the set $K = \{i \mid \varphi_i(i)\downarrow\}$. In Rice's Theorem, for example, we showed that the set I is not recursive by constructing a total computable function f such that for all $i \in \mathbf{N}$,

$$i \in K \Leftrightarrow f(i) \in I.$$

We describe this state of affairs by saying that K *is reducible to* I *via* f, meaning that f effectively transforms the membership problem for K (i.e., the problem whether $i \in K$ or $i \notin K$ for an arbitrary i) to the membership problem for I. This is equivalent to saying, in the terminology of Section 4.3, that "the function f reduces the solvability of K to the solvability of I". Hence if I were solvable then K would be too—which we know is impossible. The reduction is carried out here by the function f, but clearly other functions could perform the same transformation. If A is reducible to B via some total computable f, we shall write $A \leqslant_m B$. (We use the subscript m to distinguish this ordering from other notions of reducibility to be introduced later.) It is easy to show that $\leqslant_m$ is reflexive ($A \leqslant_m A$) and transitive ($A \leqslant_m B$ and $B \leqslant_m C \Rightarrow A \leqslant_m C$) (Exercise 1). Our first result places the set K in the $\leqslant_m$ ordering.

1 Proposition. *If A is* r.e. *then $A \leqslant_m K$.*

PROOF. Let A be an arbitrary r.e. set. Define a binary function θ by setting, for all x, y:

$$\theta(x, y) = \begin{cases} 1, & \text{if } x \in A; \\ \bot, & \text{if } x \notin A. \end{cases}$$

By Church's Thesis, θ is computable, and therefore $\theta = \varphi_e^{(2)}$ for some e. By the s-m-n Theorem, $\varphi_e^{(2)}(x, y) = \varphi_{s_1^1(e,x)}(y) = \varphi_{f(x)}(y)$. Finally, since $\varphi_{f(x)}(f(x))$ is defined if and only if $x \in A$, we conclude that $x \in A \Leftrightarrow f(x) \in K$. □

The preceding result means in effect that K is at least as *hard* to solve as any r.e. set, in the precise sense that if there were a decision procedure for membership in K, then there would also be decision procedures for all r.e. sets. Since K itself is r.e., we can say that among r.e. sets, K is a "hardest" one. K is therefore said to be *complete* in the class of r.e. sets.

Consider next the set $\overline{K} = \{i \mid \varphi_i(i)\uparrow\}$. Many of our earlier proofs showing that certain sets are not r.e. were based on a reduction from $\overline{K}$. For example, in the proofs of both Theorems 7 and 9 of the previous section we constructed a total computable function f such that for all i, $i \in \overline{K} \Leftrightarrow f(i) \in I$. This reduction from $\overline{K}$ to I forced us to conclude that I could not be r.e., because $\overline{K}$ is not. In fact, we have the following proposition.

2 Proposition. *If* $\overline{K} \leqslant_m A$ *then* A *is not* r.e.

PROOF. Since $\overline{K} = f^{-1}(A)$, $\overline{K}$ is r.e. if A is, by Proposition 4 of Section 7.2. □

The fact that $\overline{K}$ is not r.e. can be asserted in a strong way; namely, for every r.e. subset W_i of $\overline{K}$, we can effectively find a number in $\overline{K} - W_i$. Indeed, since $\overline{K} = \{i \mid i \notin W_i\}$, if $W_i \subset \overline{K}$, then $i \in \overline{K} - W_i$.

The special properties of K and $\overline{K}$ described above follow from the fact that K is a "creative" set and $\overline{K}$ is a "productive" set, two concepts we define formally in the next definition.

3 Definition

(a) A set A is *productive* if there exists a computable function θ such that for all i, if $W_i \subset A$ then $\theta(i)\downarrow$ and $\theta(i) \in A - W_i$. The function θ is called a *productive function* for A.
(b) A set B is *creative* if B is r.e. and its complement $\overline{B}$ is productive. If θ is a productive function for $\overline{B}$, we also say that θ is a *productive function* for the creative set B.

From the definition it is immediate that no productive set can be r.e. In particular, a productive set is neither finite nor cofinite.

The simplest example of a productive set is $\overline{K} = \{i \mid \varphi_i(i)\uparrow\}$, for which the identity function is a productive function (why?). Since $K = \{i \mid \varphi_i(i)\downarrow\}$ is r.e. and its complement is productive, K is creative. But productive sets do not, in general, have creative complements. For example, $\{i \mid \varphi_i \text{ nontotal}\}$ is not r.e., but its complement *is* productive, as the next example shows.

4 Example. Let $T = \{i \mid \varphi_i \text{ is total}\}$, which we already know is not r.e., by Proposition 2 of Section 4.3. We now show that T is productive by an argument that combines both diagonalization and dovetailing.

We give the argument informally. Let m be an index such that $W_m \subset T$. We start enumerating $\mathrm{DOM}(\varphi_m) = W_m$ by dovetailing. Simultaneously we construct a function $f: \mathbf{N} \to \mathbf{N}$ as follows. If at the xth step of the dovetailing procedure we do not obtain a new member of W_m, we set $f(x) = 0$; if we do obtain a member of W_m, say n, we set $f(x) = \varphi_n(x) + 1$. Since each such n belongs to T, φ_n is total so that f is also total (justifying its Roman name).

The function f is intuitively algorithmic, hence computable by a **while**-program, by Church's Thesis. Let program e compute f, i.e., $f = \varphi_e$. Since f is total, $e \in T$. If $e \in W_m$, then at some step x of the dovetailing procedure, the number e will be obtained. But then $\varphi_e(x) = f(x) = \varphi_e(x) + 1$, which is a contradiction. We are forced to conclude that $e \in T - W_m$.

We have described an algorithm to go uniformly from an index m such that $W_m \subset T$ to an index $e \in T - W_m$. Hence, by Church's Thesis, there is a computable function θ such that for all m, $\theta(m) \in T - W_m$ whenever $W_m \subset T$, as desired.

Note: We do not claim that it is possible to determine whether $W_m \subset T$ for an arbitrary m—in fact, this is not possible. And we do not claim that θ is defined only for $W_m \subset T$. Indeed we shall see later that θ may be defined for some m for which $W_m \not\subset T$ (see Lemma 10).

5 Proposition. *Every productive set contains an infinite* r.e. *subset.*

PROOF. Let A be an arbitrary productive set with productive function θ. The desired r.e. subset can be constructed in the following way. Let i_0 be an index for the empty set, i.e., $W_{i_0} = \emptyset$. Since $W_{i_0} \subset A$, $\theta(i_0) \in A - W_{i_0} = A$.

We next find an index, say i_1, for $\{\theta(i_0)\} \subset A$, i.e., $W_{i_1} = \{\theta(i_0)\}$. Since $W_{i_1} \subset A$, $\theta(i_1) \in A - W_{i_1}$. Then we find an index, say i_2, for $\{\theta(i_0), \theta(i_1)\} \subset A$, i.e., $W_{i_2} = \{\theta(i_0), \theta(i_1)\}$. Since $W_{i_2} \subset A$, $\theta(i_2) \in A - W_{i_2}$. Continuing this process, we obtain the infinite effective list

$$\theta(i_0), \theta(i_1), \theta(i_2), \ldots, \theta(i_n), \ldots$$

whose members are distinct and all contained in A. This completes the proof. □

6 Theorem. *If A is productive and $A \leqslant_m B$, then B is productive.*

PROOF. The hypothesis says that there is a total computable function f such that for all x, $x \in A \Leftrightarrow f(x) \in B$. Let θ be the productive function for A.

The reducibility of A to B via f can be restated as $A = f^{-1}(B)$. Hence, for all i, if $W_i \subset B$ then $f^{-1}(W_i) \subset A$. By Proposition 4 of Section 7.2, there is a total computable function g such that for all i, $f^{-1}(W_i) = W_{g(i)}$.

We now have:

(1) $W_i \subset B \Rightarrow W_{g(i)} \subset A \Rightarrow \theta \circ g(i) \in A - W_{g(i)}$.

On the other hand, we also have:

(2) $\theta \circ g(i) \notin W_{g(i)} \Leftrightarrow f \circ \theta \circ g(i) \notin W_i$, and
(3) $\theta \circ g(i) \in A \Leftrightarrow f \circ \theta \circ g(i) \in B$.

Putting (1), (2), and (3) together we obtain

$$W_i \subset B \Rightarrow f \circ \theta \circ g(i) \in B - W_i .$$

This shows that B is productive with productive function $f \circ \theta \circ g$. □

7 Corollary. *Let $I \subset \mathbf{N}$ be an index set which respects functions. If I satisfies the hypotheses of Theorem 7 or Theorem 9 of the preceding section, then I is productive.*

PROOF. As pointed out in Proposition 2, I was shown not to be r.e. by reducing $\bar{K}$ to I. Since $\bar{K}$ is productive, I is also productive, by the theorem. □

An immediate consequence of this result is that all the sets shown not to be r.e. in Examples 8 and 10 of the preceding section are also productive. Are all non-r.e. sets productive? We shall soon see that this is not the case.

We first turn out attention to creative sets and establish two further applications of the preceding result.

8 Corollary. *Suppose A is creative and B is* r.e. *If $A \leqslant_m B$ then B is creative.*

PROOF. Exercise 2. □

9 Corollary. *Let $I \subset \mathbf{N}$ be an index set which respects functions. If I is r.e. and $\emptyset \neq I \neq \mathbf{N}$, then I is creative.*

PROOF. We can assume that I does not contain the indices for the empty function ν. Indeed, if $\{i \mid \varphi_i = \nu\} \subset I$ and $I \neq \mathbf{N}$, then I could not be r.e., by Theorem 7 of the preceding section.

The rest of the proof follows the proof of Rice's Theorem, and shows that K is reducible to I. Finally, since K is creative, it follows that I is creative by Corollary 8. The details are left to the reader (Exercise 2). □

A consequence of Corollary 9 is that all the r.e. nonrecursive sets encountered in the book so far are creative. Although we may argue that all "natural" examples of r.e. nonrecursive sets are creative, there are nonetheless r.e. sets which are neither recursive nor creative (the simple sets of Theorem 13).

The proof of the next theorem is made easier by the following lemma, which is also of independent interest.

10 Lemma. *If set A is productive, then there is a total computable function $p : \mathbf{N} \to \mathbf{N}$ such that A is productive with productive function p.*

PROOF. Let A be productive with productive function θ, and let $\theta = \varphi_e$ for some e. Define the program-rewriting (total computable) function $g: \mathbf{N}^2 \to \mathbf{N}$ as follows:

$$\mathbf{P}_{g(i,j)} = \textbf{begin } X2 := \Phi(e, j); X2 := \Phi(i, X1) \textbf{ end}$$

This program first checks whether $\theta(j) = \varphi_e(j)$ is defined; if and when $\theta(j)\downarrow$, it goes on to compute $\varphi_i(n)$ where n is the input value stored in $X1$. Hence

$$W_{g(i,j)} = \begin{cases} W_i, & \text{if } \theta(j)\downarrow; \\ \emptyset, & \text{otherwise.} \end{cases}$$

By the extended Recursion Theorem (as stated in Exercise 10 of Section 6.1), there is a total computable function f such that for all i, $W_{f(i)} = W_{g(i,f(i))}$. Then

$$W_{f(i)} = W_{g(i,f(i))} = \begin{cases} W_i, & \text{if } \theta \circ f(i)\downarrow; \\ \emptyset, & \text{if } \theta \circ f(i)\uparrow. \end{cases}$$

Suppose now that $\theta \circ f(i)$ is undefined. Then $W_{f(i)} = \emptyset$, and since $\emptyset \subset A$, $\theta \circ f(i)$ must be defined (by productiveness)—which is a contradiction. Hence $\theta \circ f(i)$ is defined for all i, so that $W_{f(i)} = W_i$ for all i. We now have

$$W_i \subset A \Rightarrow W_{f(i)} \subset A \Rightarrow \theta \circ f(i) \in A - W_{f(i)} \Rightarrow \theta \circ f(i) \in A - W_i.$$

The desired productive function p for A is therefore $\theta \circ f$. □

We can now state and prove an important characterization of creative sets.

11 Theorem. *A set B is creative if and only if B is* r.e. *and $A \leqslant_m B$ for all* r.e. *sets A.*

PROOF. Suppose first that B is r.e., and every r.e. set is reducible to B. In particular K is reducible to B. Since K is creative, B is creative by Corollary 8.

For the converse, suppose B is creative. If we can show that $K \leqslant_m B$, then every r.e. set will be reducible to B, by Proposition 1.

Since B is creative, $\bar{B}$ is productive. By Lemma 10, we can take the productive function for $\bar{B}$ to be a total computable function p. We next define a program-rewriting (total computable) function $g: \mathbf{N}^2 \to \mathbf{N}$ by putting for all i, j:

$$\begin{aligned} \mathbf{P}_{g(i,j)} = \textbf{begin } & X2 := \Phi(i, i); X2 := \Phi(e, j); \\ & \quad \textbf{while } X1 \neq X2 \textbf{ do begin end} \\ \textbf{end} & \end{aligned}$$

where e is an index for p, i.e., $p = \varphi_e$. This program first checks whether

$\varphi_i(i)\downarrow$, i.e., whether $i \in K$; if and when $\varphi_i(i)$ converges, it goes on to compute $p(j) = \varphi_e(j)$, and then compare $p(j)$ with the input value stored in $X1$. Hence

$$W_{g(i,j)} = \begin{cases} \{p(j)\}, & \text{if } i \in K; \\ \emptyset, & \text{otherwise.} \end{cases}$$

By the extended form of the Recursion Theorem (Exercise 10 of Section 6.1), there is a total computable function $f: \mathbf{N} \to \mathbf{N}$ such that for all i:

$$W_{f(i)} = W_{g(i,f(i))} = \begin{cases} \{p \circ f(i)\}, & \text{if } i \in K; \\ \emptyset, & \text{otherwise.} \end{cases}$$

Hence,

$$i \in K \Rightarrow p \circ f(i) \in W_{f(i)} \Rightarrow W_{f(i)} \not\subset \bar{B} \Rightarrow p \circ f(i) \in B,$$

where the second implication follows from the fact that p is productive for $\bar{B}$. Likewise

$$i \notin K \Rightarrow W_{f(i)} = \emptyset \subset \bar{B} \Rightarrow p \circ f(i) \in \bar{B}.$$

We conclude that for all i

$$i \in K \Leftrightarrow p \circ f(i) \in B,$$

which means that K is reducible to B via $p \circ f$, as desired. □

One final, historical note on creative sets. It is possible to show that THM($\mathcal{P}$), the theorems of some recursive proof system as described in Section 7.3, is a creative set (Exercise 11).

This is the source of the name "creative", which originated with the American logican Post. The ability to produce a constructive counterexample to the claim: "This r.e. set exhausts the true sentences of arithmetic" was viewed by Post as a particularly vital feature of mathematics, and in this light he proposed the term *creative* set.

Simple and Immune Sets

We have not yet settled the question of what (if anything) lies between the recursive sets and the creative sets. That is, do r.e. sets exist which are neither recursive nor creative? Post noted that a creative set must in a sense have a very rich complement (a productive set), and proceeded to construct sets with sparse complements. To this end, he introduced the concept of a "simple" set. We start here with the dual concept of an "immune" set.

12 Definition

(a) A set A is *immune* if it is infinite and contains no infinite r.e. subset.
(b) A set B is *simple* if it is r.e. and its complement is immune.

From this definition, it is immediate that an immune set (if one exists at all) is not r.e., nor is it productive by Proposition 5. This implies that if simple sets exist, they will be neither recursive nor creative.

The existence of simple sets was established by Post. The special construction used in the proof of the next result is due to him.

13 Theorem. *Simple sets exist.*

PROOF. To show that an r.e. set B is simple, it suffices to show that: (i) B and $\overline{B}$ are infinite, and (ii) every infinite r.e. set intersects B (so that no infinite r.e. set is contained in $\overline{B}$).

To construct such a set B, consider a fixed dovetailing procedure for enumerating all r.e. sets. For each n, we put into B the first discovered (if any) member of W_n which is larger than $2n$. The resulting set B is effectively generated. In addition, for every n, at most n numbers out of $0, 1, \ldots, 2n$ can occur in B, so that both B and $\overline{B}$ are infinite. Finally, if W_n is an infinite r.e. set, then there are numbers larger than $2n$ in W_n, so that $B \cap W_n \neq \emptyset$. This completes the proof. □

As we set out to do, we have now established the existence of nonrecursive r.e. sets which are not *complete*, in the sense explained after Proposition 1. In particular, the nonrecursiveness of the simple sets cannot be proved by a reduction from K, in contrast to all the nonrecursive r.e. sets encountered before.

The existence of simple sets implies the existence of immune sets. But are there immune sets whose complements are also immune? (We know from Exercise 6 that there are productive sets whose complements are productive.) Using a novel construction, the following theorem gives a positive answer.

14 Theorem. *There are sets A such that both A and $\overline{A}$ are immune.*

PROOF. It suffices to show that both A and $\overline{A}$ intersect every infinite r.e. set. In fact we can prove the existence of $2^{\aleph_0}$ such sets A.

Let $n_0, n_1, n_2, \ldots$ be the members of $\{n \mid W_n \text{ is infinite}\}$ in increasing order. (This, of course, cannot be an effective enumeration.) We define inductively a sequence of two-element sets $\{a_0, b_0\}, \{a_1, b_1\}, \{a_2, b_2\}, \ldots$. First we set

$$\{a_0, b_0\} = \text{the two smallest numbers in } W_{n_0},$$

with $a_0 < b_0$. Assuming $\{a_i, b_i\}$ has been defined, we put

$$\{a_{i+1}, b_{i+1}\} = \text{the two smallest numbers of } W_{n_{i+1}} \text{ which are larger than both } a_i \text{ and } b_i,$$

with $a_{i+1} < b_{i+1}$. If we construct the set A by choosing one number from each two-element set in this infinite sequence, A will intersect every infinite r.e. set—and therefore will be immune. Since $\bar{A}$ will also contain one element from each two-element set (in addition to other numbers—why?), $\bar{A}$ will be immune too. Clearly there are $2^{\aleph_0}$ different ways of choosing A and $\bar{A}$, as claimed. □

We have now concluded our brief classification of unsolvable sets. Let us recapitulate the results, starting with the "most" unsolvable sets:

(1) There exist productive sets with productive complements (natural example: $\{n \mid \varphi_n \text{ is total}\}$), and immune sets with immune complements.
(2) There exist productive sets with r.e. complements (natural example: $\bar{K}$), and immune sets with r.e. complements.
(3) There exist creative sets, which are the "hardest" to solve among r.e. sets (natural example: K).
(4) There exist simple sets, which are of intermediate degree of unsolvability between recursive sets and creative sets.

What's beyond these? There is plenty. The interested reader may get a glimpse in Rogers (1967), where hypersimple, hyper-hypersimple, hyperimmune, hyper-hyperimmune, maximal, cohesive, . . . , and a profusion of other sets is examined.

We close this section by indicating other directions in the study of reducibility.

Different forms of reducibility

Let us review the concept of reducibility as it has been used in the book up to this point.[2] Suppose A and B are arbitrary sets, and $f: \mathbf{N} \to \mathbf{N}$ is a total computable function. Whenever we could assert that for all $n \in \mathbf{N}$

$$n \in A \Leftrightarrow f(n) \in B, \qquad (\$)$$

we conclude that

if we have a decision procedure for membership in B then we have a decision procedure for membership in A. (\$\$)

We summarized this situation by saying "A is reducible to B via f". The only requirement on f so far has been that it be a total computable function. Now we ask: If we know more about f, what more can we learn about the reduction from A to B? What if, for example, f is one-to-one, or

[2] Although our first proof of the nonrecursiveness of the set $\{n \mid \varphi_n \text{ total}\}$ in Theorem 4 of Section 4.3 was called a "reduction" (which showed that if $\{n \mid \varphi_n \text{ total}\}$ were decidable then there would be an effective enumeration of $\{n \mid \varphi_n \text{ total}\}$), that reduction was not of the kind discussed here—which reduces the membership problem for a set A to the membership problem for a set B. All other reductions discussed in the text can be put in the form (\$) considered here.

f is particularly easy to compute? The different properties f may have will induce different forms of reducibility.

15 Definition. If set A is reducible to set B via a one-to-one total computable function, we say that A is *one-one reducible* (abbreviated, 1-*reducible*) to B, and we write $A \leqslant_1 B$. The "m" of our standard $\leqslant_m$ ordering stands for "many-to-one". This kind of reducibility is called "m-reducibility".

We say that 1-reducibility is *stronger* than m-reducibility because if a set A is 1-reducible to a set B then A is also m-reducible to B, and because the converse (it can be shown) is not necessarily true.

Clearly 1-reducibility and m-reducibility each induces an ordering on sets of natural numbers. If sets A and B are 1-reducible (respectively, m-reducible) to each other, then A and B are said to be 1-*equivalent* (respectively, *m-equivalent*). Equivalence classes of sets relative to 1-equivalence (respectively, m-equivalence) are called 1-*degrees* (respectively, *m-degrees*) *of unsolvability*.

Relative to 1-reducibility and m-reducibility we have the notions of 1-*completeness* and *m-completeness*, respectively. An r.e. set B is 1-complete (respectively, m-complete) if every r.e. set is 1-reducible (respectively, m-reducible) to B. Although 1-reducibility induces a finer ordering than m-reducibility on degrees of unsolvability, it can be proven that 1-completeness is equivalent to m-completeness. We can thus restate Theorem 11 as: *Set B is creative* $\Leftrightarrow$ *B is m-complete* $\Leftrightarrow$ *B is* 1-*complete*.

In our earlier discussions of reducibilities we tied the notion of reduction to a decision procedure, as in ($) and ($$). But the decision procedure obtained was somewhat restrictive: If A is reducible to B, then the query "$x \in A$?" must be decided on the basis of a *single* query "$f(x) \in B$?"

More generally, we would like to link queries such as "$x \in A$?" to less restrictive algorithms which may use a decision procedure for B repeatedly as a subprogram. How do we formulate this kind of reducibility?

16 Definition. We define a **while**-program *with an oracle for B* to be a **while**-program which may also include **while** statements of the form

$$\textbf{while } X \in B \textbf{ do } \mathbb{S} \quad \text{or} \quad \textbf{while } X \notin B \textbf{ do } \mathbb{S}.$$

An "oracle for B" may be pictured as an external agent that will always supply the correct answer to questions of the form "$n \in B$?" or "$n \notin B$?", for every $n \in \mathbf{N}$. Note that the external agent is not necessarily identified with an algorithm; this allows for the possibility that B may not be recursive.

Recall that a decision procedure for membership in the set A is none other than a (necessarily total) **while**-program that computes the characteristic function χ_A of A.

17 Definition. A set A is *Turing reducible* (abbreviated *T-reducible*) to a set B if there is a **while**-program with an oracle for B which computes χ_A. We will write $A \leqslant_T B$ to mean "A is Turing reducible to B".

This concept of reducibility gives a precise meaning for the notion "A is recursive relative to B" or, more succinctly, "A is recursive in B".

It is not difficult to check that m-reducibility implies T-reducibility, while the converse implication fails (Exercise 8); that is, m-reducibility is stronger than T-reducibility. Let us call an r.e. set B *T-complete* if every r.e. set is T-reducible to B. Then every m-complete set is T-complete, but the converse if false: There are T-complete sets which are not m-complete.

Another kind of reducibility is "polynomial reducibility", which plays an important role in the classification of recursive sets and "subrecursive hierarchies". [See, for example, Machtey and Young (1978), Ladner (1975), and Kozen (1980).] If

$$n \in A \Leftrightarrow f(n) \in B$$

where f is total computable in polynomial time, we say that "A is many-one reducible to B in polynomial time". This means that the reduction function f sending $n \in A$ to its image in B must be "fast-running", i.e., must take no more than a polynomial number of steps as a function of the input n. Hence, if the decision procedure for B itself runs in polynomial time (as a function of the input), then we also have a polynomial-time decision procedure for A.

The restricted reducibility we have just defined is called *polynomial m-reducibility*. We can also define the notion of *polynomial T-reducibility* by restricting **while**-programs with oracles to operate within polynomial time, i.e., within a polynomial number of steps (as a function of the input). Polynomial m-reducibility is stronger than polynomial T-reducibility.

Post's Problem

We conclude with a historical note. In 1944, Emil Post raised the question of whether there are r.e. sets of intermediate T-degree between recursive sets and T-complete sets—just as there are r.e. sets (the simple sets) which are of intermediate m-degree between recursive sets and m-complete sets (the creative sets). This question became known as *Post's Problem* and played an important role in the early development of the theory of degrees of unsolvability. Post's Problem is not settled by the existence of simple sets which, it turns out, can be T-complete (but not m-complete, of course). Post had hoped that by placing even more stringent requirements on the complement of an r.e. set A, for which he invented the concepts of "hyperimmune" and "hyper-hyperimmune" sets, he could force A to be of intermediate T-degree. This approach turned out to be unsuccessful.

Then in 1956, two years after Post's death, the problem was solved simultaneously and independently by Friedberg (in his senior thesis at Harvard) and by Muchnik (in the Soviet Union). They showed that r.e. sets of intermediate T-degree between recursive sets and T-complete sets do exist, and in great profusion. Their proof technique, now known as the *priority method*, is of fundamental importance in the theory of degrees of unsolvability.

Exercises for Section 8.2

1. Prove that $\leqslant_m$ is reflexive and transitive.

2. (a) Prove Corollary 8.
   (b) Fill in the details of Corollary 9.

3. (a) Prove that every infinite r.e. set is the disjoint union of a creative set and a productive set.
   (b) Use part (a) to prove that every productive set is the disjoint union of a creative set and a productive set.
   (*Hint*: You need to use Proposition 5 also.)

4. (Rogers) Show that every one-to-one total computable function is the productive function for some creative set. (*Hint*: Try $\{f(i) \mid f(i) \in W_i\}$.)

5. Using the proof of Theorem 11 as a guide, prove directly, i.e., without invoking Proposition 1, that if set B is creative then every r.e. set is reducible to B.

6. (a) Prove that both $\{n \mid \varphi_n \text{ total}\}$ and $\{n \mid \varphi_n \text{ not total}\}$ are productive sets.
   (b) Give two other "natural" examples of productive sets with productive complements. Justify your answer.

7. (a) Exhibit two simple sets whose union is $\mathbf{N}$.
   (b) If A is a simple set and B an arbitrary r.e. set, prove that $A \cup B$ is simple or cofinite.

8. (a) Prove that if A is m-reducible to B then A is T-reducible to B.
   (b) Prove that the converse of the implication in (a) is not generally true. (*Hint*: $\overline{K}$ is T-reducible to K.)
   (c) Prove that every m-complete set is also T-complete.

9. Consider the set M of "minimal" indices:

   $$M = \{i \mid \text{for all } j \text{ if } \varphi_j = \varphi_i \text{ then } j \geqslant i\}.$$

   M can be though of as the set of "shortest" **while**-programs, one for every unary computable function (see also Exercise 9 of Section 4.3).

   (a) Show that every infinite r.e. set intersects $\overline{M}$, the complement of M. (*Hint*: If f is a total computable function with infinite range, then the function $\theta(i) = f(\mu z[f(z) > i])$ is total computable such that $\theta(i) > i$ for all $i \in \mathbf{N}$. Consider $\varphi_{\theta(i)}(x)$ and apply the Recursion Theorem.)
   (b) Exhibit an infinite r.e. subset of $\overline{M}$.

(c) Prove that $\overline{M}$ is not r.e. (*Hint*: Let i_0 be the minimal index for the empty function. Argue that for all i,

$$i \in \{n \mid \varphi_n \equiv \perp\} \Leftrightarrow (\forall j < i)\big[\, j = i_0 \text{ or } \varphi_j \not\equiv \perp \text{ or } j \in \overline{M}\,\big]$$

so that if $\overline{M}$ was r.e. then $\{n \mid \varphi_n \equiv \perp\}$ would be r.e. too.)

(d) Conclude that M is an immune set whose complement is neither immune nor r.e.

10. Prove that if A is recursive and $A \cap B$ is productive then B is productive.

11. (This result assumes knowledge of Section 7.3.) Assume that $\mathcal{P}$ is a recursive proof system with the property that if $n \in W_m$, then $\vdash_{\mathcal{P}} W(\bar{n}, \bar{m})$. (This is a reasonable assumption. It asserts that $\mathcal{P}$ can "track" computations by **while**-programs.)

(a) Prove that TR is a productive set. (*Hint*: Use Exercise 10.)

(b) Show that $K \leqslant_m \text{THM}(\mathcal{P})$, and conclude that $\text{THM}(\mathcal{P})$ is creative.

CHAPTER 9

Alternative Approaches to Computability

In the previous chapters we have developed a theory of computability based on **while**-programs. The legitimacy of this programming language approach has rested on the belief, embodied in Church's Thesis, that the class of computable functions is model invariant: The **while**-program computable functions include the function classes that are definable by any other system, so long as that system matches our informal notion of algorithmic specification.

In this chapter we present strong evidence in support of our commitment to Church's Thesis. We consider two other formulations of computability, one due to Turing, the other due to Kleene, which differ dramatically from the **while**-program formulation, but which nevertheless define the same class of computable functions. We also briefly consider several other special computing formalisms for string processing, and we identify certain undecidable properties for these systems.

9.1 The Turing Characterization

The Turing-machine formalism was one of the first systems proposed as a model for algorithmic specification. Although Turing machines seem to resemble a primitive assembly language on a real digital computer, their formulation by the British mathematician Alan Turing in 1936 predates the computer age by almost a decade [Turing (1936)]. In this section we describe Turing's formulation of computability and demonstrate that it qualifies as an acceptable programming system in the sense of Section 6.2.

TURING MACHINES

A Turing machine is a finite-state control equipped with an external storage device. This storage mechanism may be thought of as a finite tape that can be extended indefinitely in both directions:

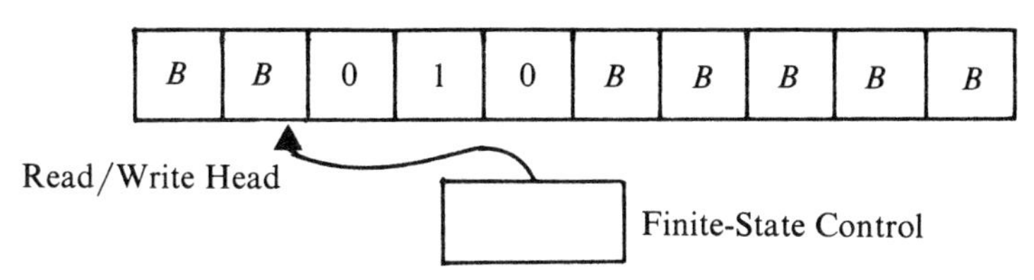

Figure 1 A Turing machine: The tape can be extended indefinitely by adding B's at either end.

As shown in Figure 1, the tape is divided into *squares*. Each square of the tape may be blank, or may hold any one symbol from a specified finite tape alphabet Σ. For convenience, a special symbol (here, the letter "B") not in the alphabet Σ is reserved to denote a blank square on the tape. Thus, in Figure 1, all squares are blank except for three, which hold the symbols 0, 1, and 0. The finite-state control is coupled to the tape through a read/write head. At any given instant, the head will be scanning one square of the tape and the finite-state control will be in one state. Depending on this state and the symbol in the square scanned by the tape head, the machine will, in one step, do all of the following:

(1) Enter a new state of the finite-state control;
(2) overwrite a symbol on the scanned square (it is possible to overwrite the same symbol and so leave it unchanged, or also overwrite it by B and so "erase" the symbol);
(3) either shift the head left one square or right one square, or not shift it at all.

Since the tape alphabet is finite and the finite-state control has finitely many states, the operation of a Turing machine can be completely described by a finite set of *quintuples* of the form:

(old state, symbol scanned, new state, symbol written, direction of motion)

1 Example. The following set of quintuples describes a Turing machine with four states q_0, q_1, q_2, q_H, which operates over the tape alphabet

$\Sigma = \{1\}$:

$$\begin{array}{lllll}(q_0 & B & q_1 & B & R)\\(q_1 & 1 & q_1 & 1 & R)\\(q_1 & B & q_2 & 1 & L)\\(q_2 & 1 & q_2 & 1 & L)\\(q_0 & 1 & q_H & 1 & N)\\(q_2 & B & q_H & B & N)\end{array}$$

In these quintuples, "R" denotes a rightward motion of the tape head, and "L" a leftward motion. An "N" denotes no move.

We can give an equivalent representation of the quintuples in Example 1 in terms of a *state table*, as follows:

Old state	Symbol scanned 1	Symbol scanned B
q_0	q_H 1 N	q_1 B R
q_1	q_1 1 R	q_2 1 L
q_2	q_2 1 L	q_H B N
q_H	—	—

In order to make the behavior of a machine deterministic, we require that whenever two quintuples have the same (old state, symbol scanned) combinations, then the two quintuples are identical. Summarizing:

2 Definition. A Turing machine Z is specified by a quintuple (Q, X, q_0, q_H, δ) where

Q is a finite set of *states*;
$X = \Sigma \cup \{B\}$ is a finite set of *tape-symbols*, where $B \notin \Sigma$ is the *blank*;
q_0 in Q is the *initial state*;
q_H in Q is the *halt state*;
$\delta : Q \times X \to Q \times X \times \{L, N, R\}$ is the (partial) *transition function*, with $\delta(q_H, x)$ undefined for each x in X.

As we have seen, we may represent δ by a list of quintuples, with $q\ x\ q'\ x'\ d$ in the list just in case $\delta(q, x) = (q', x', d)$.

If we start Z in state q_0 scanning a given square of its tape, the *computation* that ensues will be uniquely defined by the quintuples of δ. We further make the convention that two tapes (which are elements of the set X^* of finite strings over the alphabet X of tape symbols) are equivalent if they differ by splicing strings of B's on either end. If the machine tries to move off the end of the tape, we simply splice an extra B on that end of the tape. In this way, Z will start from a finite tape configuration (the input), and if its computation *terminates* after finitely many steps—by reaching q_H —it will leave a finite tape configuration (the output).

Going back to Example 1, it is easily checked that whenever the Turing machine is given a tape configuration of the form

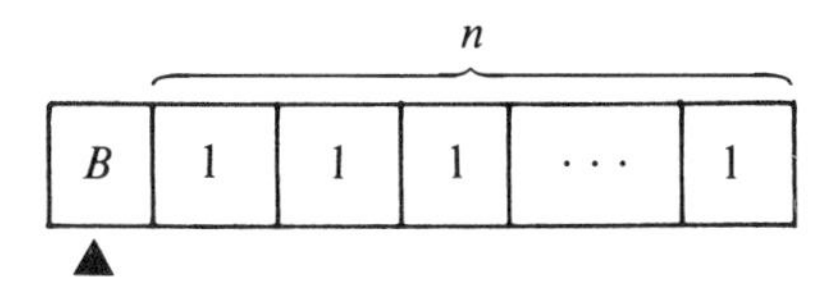

i.e., an everywhere-blank tape except for a block of n consecutive 1's, then its computation with starting state q_0, and with the tape head initially positioned on the first B to the left of the block of 1's, may be described as follows:

(1) The head keeps moving rightward leaving the 1's unchanged;
(2) upon encountering the first blank square to the right, a 1 is printed and the head starts moving leftward;
(3) upon encountering the first blank to the left, the computation terminates, leaving a block of $(n + 1)$ consecutive 1's on the tape.

Turing-machine semantics

By what we have just seen, Turing machines are primitive theoretical machines. They operate locally, shifting only at most one square left or right when a particular instruction is executed. Turing machines appear to be general symbol-processing devices rather than machines that compute number-theoretic partial functions. Nevertheless it is possible to interpret Turing machines as computers for such functions, and we next establish these semantics conventions with an eye toward showing that Turing machines are in fact as powerful as **while**-programs.

First of all, let us agree to write natural numbers in unary notation. That is, we represent the natural number n by a block of $(n + 1)$ consecutive 1's. For convenience, we refer to the unary representation of n by the symbol $\bar{n}$. Thus $\bar{0}$ is 1, $\bar{1}$ is 11, $\bar{2}$ is 111, etc.

Traditionally, Turing machine M computes a function

$$\tau_M : \mathbf{N}^k \rightarrow \mathbf{N}$$

according to the following rules:

(1) To calculate $\tau_M(m_1, \ldots, m_k)$, start M on initial configuration

$$\underset{\blacktriangle}{B} \quad \overline{m}_1 \quad B \quad \overline{m}_2 \quad B \quad \ldots \quad B \quad \overline{m}_k .$$

(2) If and when M halts, count up the total number of 1's on the tape, interpret this number in unary notation, and take this value to be the value of $\tau_M(m_1, \ldots, m_k)$.

However, we wish to modify these conventions slightly, so as to simplify the job of showing that Turing machines can simulate **while**-programs. Accordingly for any M and any integers $k, l \geqslant 1$, we define a semantics function

$$\tau_M : \mathbf{N}^k \to \mathbf{N}^l$$

as follows:

(1) To calculate $\tau_M(m_1, \ldots, m_k)$, start M on initial configuration

$$\underset{\blacktriangle}{B} \quad \overline{m}_1 \quad B \quad \overline{m}_2 \quad \ldots \quad B \quad \overline{m}_k .$$

(2) If and when M halts, the halting configuration must look like

$$\underset{\substack{\blacktriangle \\ q_H}}{B} \quad \overline{n}_1 \quad B \quad \overline{n}_2 \quad \ldots \quad B \quad \overline{n}_l \quad B \quad \ldots$$

where only blank squares may appear to the left of the head. (Other symbols besides blanks may appear to the right of the blank right boundary marker of the lth block.) If these conditions are met, then $\tau_M(m_1, \ldots, m_k) = (n_1, \ldots, n_l)$. Otherwise the value of $\tau_M(m_1, \ldots, m_k)$ is undefined.

Although the initial and final tape patterns are required to be over the alphabet $\{1, B\}$, we nevertheless permit Turing machines to use auxiliary symbols in the course of computations (see Exercise 2).

Our next example illustrates more concretely how a Turing machine may be viewed as a crude vector processor. Roughly speaking, the Turing machine presented performs the operation "$X1 := succ(X3)$". The hard work comes in leaving the values of $X2$ and $X3$ unaltered.

3 Example. The Turing machine given below performs the following sequence of operations on an input of the form

$$\underset{\blacktriangle}{B} \quad \overline{n}_1 \quad B \quad \overline{n}_2 \quad B \quad \overline{n}_3$$

(a) $\overline{n}_1$ is erased, and the boundary blank between $\overline{n}_1$ and $\overline{n}_2$ is replaced with a #:

$$\underset{\blacktriangle}{\#} \quad \overline{n}_2 \quad B \quad \overline{n}_3$$

(b) The $\overline{n}_3$ block is converted to a block of $*$'s followed by an end marker $+$:

$$\# \quad \overline{n}_2 \quad B \quad \underbrace{* \cdots *}_{n_3+1} \quad \underset{\blacktriangle}{+}$$

(c) One by one the $*$'s are converted back to 1's, and each time this happens, 1's are copied to the left of the #:

$$\underbrace{1 \cdots 1}_{j} \quad \# \quad \overline{n}_2 \quad B \quad * \cdots * \quad \underbrace{1 \cdots 1}_{j} \quad \underset{\blacktriangle}{+}$$

(d) When all the $*$'s are converted, the $+$ is converted to a blank, the $\#$ is converted to a blank, and one additional 1 is copied at the far left:

$$\begin{array}{ccccccc} B & \overline{n_3+1} & B & \bar{n}_2 & B & \bar{n}_3 & B \\ \blacktriangle & & & & & & \\ q_H & & & & & & \end{array}$$

We give the full machine below:

q_0	1	B	R	q_0	(blanks the first block)
q_0	B	$\#$	R	q_1	(leaves marker)
q_1	1	1	R	q_1	(moves to block 3)
q_1	B	B	R	q_2	
q_2	1	$*$	R	q_2	(changes 1's to $*$'s)
q_2	B	$+$	L	q_3	(leaves right marker)
q_3	$*$	1	L	q_4	(converts right-most $*$ to 1)
q_4	$*$	$*$	L	q_4	(proceeds back to block 1)
q_4	1	1	L	q_4	
q_4	B	B	L	q_4	
q_4	$\#$	$\#$	L	q_5	(enters last block)
q_5	1	1	L	q_5	(moves to left end of block)
q_5	B	1	R	q_6	(marks a left-most 1)
q_6	1	1	R	q_6	(moves to right end of block 3)
q_6	B	B	R	q_6	
q_6	$\#$	$\#$	R	q_6	
q_6	$*$	$*$	R	q_6	
q_6	$+$	$+$	L	q_7	(turns around)
q_7	1	1	L	q_7	(finds right-most $*$, converts it to a 1, and repeats)
q_7	$*$	1	L	q_4	
q_7	B	B	R	q_8	(all block 3 $*$'s converted; head moves right)
q_8	1	1	R	q_8	
q_8	$+$	B	L	q_9	(right end marker erased)
q_9	1	1	L	q_9	(head moves left until $\#$ found)
q_9	B	B	L	q_9	
q_9	$\#$	B	L	q_{10}	
q_{10}	1	1	L	q_{10}	(left-most block of 1's traversed)
q_{10}	B	1	L	q_H	(extra 1 inserted)

Notice that our semantics conventions bear an obvious resemblance to **while**-program semantics conventions, with k input blocks corresponding to the association of **while**-program inputs with program variables. We will exploit this resemblance soon when we prove our main equivalence theorem.

4 Definition. A partial function $\psi : \mathbf{N}^k \to \mathbf{N}$ is *Turing computable* if there exists a Turing machine M such that $\tau_M = \psi$.

We now demonstrate that Turing machines are an acceptable programming system in the sense of Section 6.2; i.e., Turing machines are complete, have Turing-computable universal functions of every arity, and satisfy the s-m-n property. We begin with completeness.

5 Theorem. *A partial function $\psi : \mathbf{N}^k \to \mathbf{N}$ is* **while**-*program computable if and only if it is Turing computable.*

PROOF. We prove that **while**-program computability implies Turing computability. The proof in the other direction is easier, and is requested in the exercises (Exercise 4).

Recall that a k-variable **while**-program is a vector processor, taking a k-vector as input, and producing as output another k-vector. To interpret program behavior as a function, we extract the value of variable $X1$. We now show how the vector-processor view of **while**-programs can be simulated by Turing machines. This demonstration will be sufficient to prove the theorem, since for arbitrary Turing machine M, it is easy to see that by reinterpreting $\tau_M : \mathbf{N}^k \to \mathbf{N}^k$ as a map instead of the form $\tau_M : \mathbf{N}^k \to \mathbf{N}$ we will have, in effect, extracted the value of $X1$.

Our proof that **while**-program computability implies Turing computability is based on the inductive definition of **while**-programs given in Section 2.1. As our basis step we must establish that the successor and predecessor functions and the zero assignment operations are all Turing computable. Example 1 of this section is the successor function. The predecessor machine is defined similarly, and is left as an exercise. Zero assignment:

$$Xj := 0$$

can be executed by a machine which does the following. Given the input

$$\underset{\blacktriangle}{B} \ \ \overline{m}_1 \ \ B \ \ \cdots \ \ B \ \ \overline{m}_j \ \ B \ \ \cdots \ \ B \ \ \overline{m}_k,$$

the machine M:

(1) shifts right until it finds the jth block of 1's;
(2) erases all but the first of the 1's in that block; and
(3) moves blocks $(j + 1)$ to k leftward; and
(4) shifts back to the B preceding $\overline{m}_1$.

Thus, $\tau_M : \mathbf{N}^k \rightarrow \mathbf{N}^k$ takes a k vector $(m_1, \ldots, m_j, \ldots, m_k)$ and returns $(m_1, \ldots, 0, \ldots, m_k)$ as output.

There are two induction steps. Suppose P is a k-variable **while**-program which is computed by a compound statement of the form

$$\textbf{begin } \mathbb{S}_1; \mathbb{S}_2; \ldots ; \mathbb{S}_n \textbf{ end}$$

By induction, each k-variable program $\mathbb{S}_i$ is computable by a Turing machine T_i. We must show how to construct a machine that computes P. We begin by renaming states in the T_i's so that the n state sets are disjoint. Call these machines $T_1', T_2', \ldots, T_n'$. Then we write a simple control program which guarantees that T_{i+1}' execution follows T_i' execution smoothly. So our program:

(1) Executes T_1' so that after execution, the k-ary vector result of $\mathbb{S}_1$ execution is correctly represented on the tape, with the head immediately to the left of this representation;
(2) $T_2', \ldots, T_n'$ are executed in order.

Induction step 2 requires that we demonstrate how to simulate

$$\textbf{while } Xi \neq Xj \textbf{ do } \mathbb{S}$$

given that we have in hand a machine for simulating $\mathbb{S}$, say $T_{\mathbb{S}}$. Again $\mathbb{S}$ is a k-variable **while**-program, which we treat as a vector processor taking k-vectors to k-vectors. In addition to $T_{\mathbb{S}}$ we need an auxiliary machine, $T_{\text{check}(i,j)}$, which checks to see if the ith block of 1's equals the jth block of 1's. Using these machines, the simulation proceeds as follows. First, $T_{\text{check}(i,j)}$ is executed; if this check is positive (i.e., if $Xi = Xj$), then the head is moved back to the left, and the program is terminated. Otherwise $T_{\mathbb{S}}$ is called, after which $T_{\text{check}(i,j)}$ is called again, etc. □

We have thus sketched how any **while**-program computable function may be simulated by a Turing machine. Our result is surprising because at first glance Turing machines seem so much more primitive than **while**-programs.

Next, we briefly show from first principles that for $k \geqslant 1$, k-ary Turing-computable functions possess a Turing-computable $(k + 1)$-ary universal function. First we extend the limited character code chart of Chapter 3 so as to include all elements of the quintuple vocabulary of Turing machines. Then, as in Section 3.1, we give an effective enumeration of Turing machines (see Exercise 7):

$$T_0, T_1, \ldots, T_k, \ldots .$$

By interpreting Turing-machine output as presented earlier, we can give an effective listing of the j-ary Turing computable functions,

$$\tau_0^{(j)}, \tau_1^{(j)}, \ldots, \tau_i^{(j)}, \ldots .$$

A Turing-computable $(j + 1)$-ary universal function $\tau_u^{(j+1)}$ is computed by a Turing machine T_u which, when initialized to

$$\underset{\blacktriangle}{B} \quad \bar{x} \quad B \quad \bar{y}_1 \quad B \quad \cdots \quad B \quad \bar{y}_j,$$

interprets $\bar{x}$ as a Turing machine instruction set and executes T_x on $y_1, \ldots, y_j$. That such a T_u exists is eminently believable. T_u must have at its disposal machinery to decode x and represent this decoding at a remote portion of its tape; machinery to represent T_x's tape, and machinery to fetch T_x's instructions and simulate their execution on T_u's representation of T_x's tape.

While all this is believable, the details are tedious and we omit them. The interested reader will find this construction spelled out in detail in Arbib (1969), Davis (1958), and Minsky (1967).

Thus we have:

6 Theorem. *For every $j \geqslant 1$ there exists a Turing machine T_u such that $\tau_u^{(j+1)}(e, x_1, \ldots, x_j) = \tau_e(x_1, \ldots, x_j)$.*

Finally, we have:

7 Theorem. *The s-m-n property holds for the Turing machine formalism.*

Sketch of Proof. Recall our proof of the s-m-n theorem for **while**-programs (Theorem 1 of Section 4.2): The function s_n^m in the equation

$$\varphi_{s_n^m(e, x_1, \ldots, x_m)}(y_1, \ldots, y_n) = \varphi_e(x_1, \ldots, x_m, y_1, \ldots, y_n)$$

is the mapping which takes the code for $\mathbf{P}_e$ and maps it to the code

```
begin
    block of assigment statements;
    P_e
end
```

There is a similar s-m-n equivalence for Turing machines. Intuitively it asserts that T_e when run on $B\ \bar{x}_1\ B\ \bar{x}_2\ \cdots\ \bar{x}_m\ B\ \bar{y}_1\ B\ \cdots\ B\ \bar{y}_n$ is equivalent to $T_{s_n^m(e, x_1, \ldots, x_m)}$ on initial tape

$$\underset{\blacktriangle}{B} \quad \bar{y}_1 \quad B \quad \cdots \quad B \quad \bar{y}_n \quad B.$$

$s_n^m(e, x_1, \ldots, x_m)$ codes the following sequence of Turing machine actions:

(1) Write $B\ \bar{x}_1\ B\ \cdots\ B\ \bar{x}_m$ on the tape to the left of the current head position.

(2) Move the head to the left, so that the tape looks like

$$\underset{\blacktriangle}{B} \quad \bar{x}_1 \quad B \quad \cdots \quad B \quad \bar{x}_m \quad B \quad \bar{y}_1 \quad B \quad \cdots \quad B \quad \bar{y}_n \quad B$$

(3) Execute T_e.

Each of (1), (2), and (3) are clearly actions programmable with Turing machine instructions. □

Theorems 5, 6, and 7 establish that Turing machines are an acceptable programming system.

EXERCISES FOR SECTION 9.1

1. Give a Turing machine which computes the predecessor function.
2. Prove that a Turing machine over an arbitrary finite alphabet Σ can be simulated by a machine over $\Sigma' = \{1, B\}$.
3. Give a Turing machine which performs multiplication.
4. Prove that every Turing computable function is **while**-program computable.
5. Construct Turing machine $T_{\text{check}(i,j)}$ described in the proof of Theorem 5.
6. Consider all Turing machines which move only to the right. That is, consider machines for which the δ function of Definition 2 is of the form
$$\delta : Q \times X \to Q \times X \times \{R, N\}.$$
Is this machine set an acceptable programming system?
7. Give a Gödel numbering for Turing machines.
8. Is $\{x \mid T_x \text{ halts when started on a blank tape}\}$ decidable?

9.2 The Kleene Characterization

One standard description of the addition function involves the use of a *recursive* definition and a help function, *succ*:

$$plus(x, 0) = x;$$
$$plus(x, n+1) = succ(plus(x, n)).$$

Multiplication can be specified in this way as well using *plus* as a help function:

$$times(x, 0) = x;$$
$$times(x, y+1) = plus(x, times(x, y)).$$

The ability to use recursive definitions to construct new functions from given functions is one distinctive feature of Kleene's characterization of the computable functions. Kleene's characterization was formulated in the 1930's and is related to work done around that time by Gödel and Herbrand [Kleene (1936)]. Later in this section we show how Kleene's approach can be made into an acceptable programming system.

There are two stages in Kleene's formulation. First he describes the so-called *primitive recursive* functions. This class contains a small set of starting functions, including the *successor* function, and is closed under the operations of composition and primitive recursion. The definitions of *plus* and *times* given above are examples of this latter operation.

The primitive recursive functions are a broad and interesting class of functions, and include most functions ever encountered in practical mathematics and computer science. However, all primitive recursive functions are total, and hence the class must necessarily fall short of the full class of computable functions. Ackermann's function, which is described in Exercise 13 of Section 2.3, is an example of a total computable but nonprimitive recursive function.

The second stage of the Kleene chracterization extends the class of primitive recursive functions by adding an additional operator, the μ-operator, which introduces unbounded and possibly nonterminating searches. The μ-operator applied to the class of total partial recursive functions defines the *partial recursive* functions, a class which coincides with the class of functions computable by **while**-programs.

THE PRIMITIVE RECURSIVE FUNCTIONS

1 Definition (The Primitive Recursive Functions)

(a) The following functions are primitive recursive:

1. The zero function $Z(x) \equiv 0$;
2. The successor function $succ(x) = x + 1$;
3. The projection functions $U_i^n(x_1, \ldots, x_n) = x_i$ for all i and n, $1 \leqslant i \leqslant n$.

(b) Suppose $g_1, \ldots, g_k, h$ are primitive recursive, where $h : \mathbf{N}^k \to \mathbf{N}$ and for each j, $1 \leqslant j \leqslant k$, $g_j : \mathbf{N}^n \to \mathbf{N}$. Then

$$h(g_1(x_1, \ldots, x_n), \ldots, g_k(x_1, \ldots, x_n))$$

is primitive recursive.

(c) Suppose $g : \mathbf{N}^k \to \mathbf{N}$ and $h : \mathbf{N}^{k+2} \to \mathbf{N}$ are primitive recursive. The f is the primitive recursive also, where f is defined by

$$f(0, x_1, \ldots, x_k) = g(x_1, \ldots, x_k)$$
$$f(n+1, x_1, \ldots, x_k) = h(f(n, x_1, \ldots, x_k), n, x_1, \ldots, x_k).$$

Part (a) of the above definition identifies the basis functions for the primitive recursive functions. Part (b) says that the *composition* of primitive recursive functions is primitive recursive. And part (c) formally describes the notion of *primitive recursion* presented earlier.

Notice the form of the recursive step in (c). The $(n+2)$-ary function h must be passed the value n and the parameter list $x_1, \ldots, x_k$ as well as the value of f on these arguments. So, strictly speaking, the definitions for *plus* and *times* given above are incorrect. However, it is not hard to see how to incorporate and then ignore these extra parameters so as to make these definitions strictly legal. For example, we can write *plus* as

$$plus(0, x_1) = U_1^1(x_1)$$

$$plus(n+1, x_1) = succ\big(U_1^3(plus(n, x_1), n, x_1)\big).$$

If we regard $U_1^1(x_1)$ as $g(x_1)$ and $succ(U_1^3(x_1, x_2, x_3)$ as $h(x_1, x_2, x_3)$, then *plus* is indeed primitive recursive. A similar precise description may be given for *times* (see Exercise 1). From now on we will give only semiformal primitive recursive descriptions when it is clear how to make such descriptions strictly legal.

Using the initial functions in conjunction with the two combining rules, composition and primitive recursion, it is possible to build up an impressive list of primitive recursive functions.

2 Proposition. *The following functions are primitive recursive.*

(a)
$$fact(n) = \begin{cases} 1, & \text{if } n = 0; \\ n * \cdots * 2 * 1, & \text{otherwise}. \end{cases}$$

(b)
$$sgn(n) = \begin{cases} 0, & \text{if } n = 0; \\ 1, & \text{otherwise}. \end{cases}$$

(c)
$$\overline{sgn}(n) = \begin{cases} 1, & \text{if } n = 0; \\ 0, & \text{otherwise}. \end{cases}$$

(d)
$$pred(n) = \begin{cases} n-1, & \text{if } n > 0; \\ 0, & \text{otherwise}. \end{cases}$$

(e)
$$x \dot{-} y = \begin{cases} x-y, & \text{if } x > y; \\ 0, & \text{otherwise}. \end{cases}$$

(f)
$$x < y = \begin{cases} 1, & \text{if } x < y; \\ 0, & \text{otherwise}. \end{cases}$$

(g) x^y.

(h) $|x-y|$.

PROOF.

(a) $fact(0) = 1 = succ(Z(0))$,
$fact(n + 1) = times(n, fact(n))$;
(b) $sgn(0) = 0$;
$sgn(n + 1) = succ(Z(n))$;
(c) $\overline{sgn}(0) = 1$;
$\overline{sgn}(n + 1) = Z(n)$;
(d) $pred(0) = 0$;
$pred(n + 1) = n$;
(e) $x \dot{-} 0 = x = U_1^1(x)$;
$x \dot{-} (y + 1) = pred(x \dot{-} y)$;

(f), (g), and (h) are left as exercises. □

Before proceeding further with our development of the primitive recursive functions, we include an observation that will help clarify the limits of primitive recursive descriptions.

3 Proposition. *Every primitive recursive function is total.*

PROOF. The proof is by induction on the depth of nested instances of composition and primitive recursion in a primitive recursive description. If this number is 0, then the function must be an initial function, and hence total by definition. If the theorem is true for all primitive recursive descriptions of depth n, then it is true for depth $n + 1$ as well, since the composition of total functions is total, and a function defined by primitive recursion from total functions must also be total. □

4 Proposition. *There exist non-primitive-recursive total computable functions.*

PROOF. It is possible to give a systematic listing of all primitive recursive functions of one variable (Exercise 5). Let this listing be

$$f_0, f_1, f_2, \ldots, f_n, \ldots .$$

Then by Church's thesis, $g(x) = f_x(x) + 1$ is computable. But g cannot be primitive recursive, since it would then have a primitive recursive index, say e, which would imply $f_e(e) = f_e(e) + 1$. □

Next we identify certain auxiliary constructions which preserve primitive recursiveness. By a *primitive recursive predicate* we mean a predicate with a primitive recursive characteristic function.

5 Proposition. *Suppose $Q(n)$ is a primitive recursive predicate, and suppose $f(n)$ and $g(n)$ are primitive recursive functions. Then the function*

$$r(n) = \textbf{if } Q(n) \textbf{ then } f(n) \textbf{ else } g(n)$$

that is,

$$r(n) = \begin{cases} f(n), & \text{if } Q(n) = 1, \\ g(n), & \text{if } Q(n) = 0, \end{cases}$$

is primitive recursive.

PROOF. Define $r(n)$ as follows:

$$r(n) = Q(n) * f(n) + (1 \dot{-} Q(n)) * g(n).$$

□

6 Corollary. *A boolean combination of primitive recursive predicates is primitive recursive.*

PROOF. Since **if-then-else** is a complete connective for the propositional calculus, all other connectives can be expressed in **if-then-else** form. For example, $Q(n) \vee R(n)$ may be written as

if $Q(n)$ **then** 1 **else** $R(n)$.

□

7 Corollary. *The predicates* $=$, $\neq$, *and* $\leqslant$ *are primitive recursive.*

PROOF. Exercise 6. □

We indicated earlier that the absence of unbounded searches is an important feature of the primitive recursive functions. However, if searches are suitably bounded, then primitive recursiveness is preserved. For convenience we repeat here the definition of Exercise 9 of Section 2.3.

8 Definition. Let $Q(k)$ be a predicate. The *bounded minimization operator* applied to $Q(k)$ defines a function f which we express as

$$f(n) = \mu k \leqslant n[Q(k) = 1].$$

The expression $\mu k \leqslant n[Q(k) = 1]$ asks for the argument of the first element in the sequence

$$Q(0), Q(1), \ldots, Q(n)$$

which is *true*, i.e., which has the value 1. If there is no $k \leqslant n$ such that $Q(k) = 1$, the value of $f(n)$ is 0.

The definition extends in an obvious way to bounded minimization over predicates with parameters, e.g.,

$$f(n, x_1, \ldots, x_j) = \mu k \leqslant n[Q(k, x_1, \ldots, x_j) = 1].$$

9 Proposition. *Let* $Q(n)$ *be a primitive recursive predicate, and define*

$$f(n) = \mu k \leqslant n[Q(k) = 1].$$

Then f *is primitive recursive.*

PROOF. Define f as

$$f(0) = 0,$$

$$f(n+1) = \begin{cases} n+1, & \text{if } Q(0) \neq 1, f(n) = 0, \text{ and } Q(n+1) = 1; \\ f(n), & \text{otherwise.} \end{cases}$$

□

We now return to the development of common primitive recursive functions of number theory.

10 Proposition. *The following functions are primitive recursive.*

(a)
$$\mathit{quotient}(m,n) = \begin{cases} 0, & \textit{if } n = 0; \\ \textit{the integer part of } m/n, \textit{ otherwise.} \end{cases}$$

(b)
$$\mathit{rm}(m,n) = \begin{cases} 0, & \textit{if } n = 0; \\ \textit{the remainder of } m/n, \textit{ otherwise.} \end{cases}$$

(c)
$$\mathit{power}(m,n) = \begin{cases} 1, & \textit{if } m = n^k \textit{ for some } k; \\ 0, & \textit{otherwise.} \end{cases}$$

(d)
$$\mathit{prime}(n) = \begin{cases} 1, & \textit{if } n \textit{ is a prime}; \\ 0, & \textit{otherwise.} \end{cases}$$

(e)
$$\mathit{pr}(n) = \textit{the nth prime in the sequence of primes, given that 2 is the 0th prime number.}$$

PROOF.

(a)
$$\mathit{quotient}(m,n) = \begin{cases} 0, & \text{if } n = 0; \\ \mu z \leqslant m[(z+1) * n > m], & \text{otherwise.} \end{cases}$$

(b)
$$\mathit{rm}(m,n) = \begin{cases} 0, & \text{if } n = 0; \\ m \dot{-} (n * \mathit{quotient}(m,n)), & \text{otherwise.} \end{cases}$$

(c)
$$\mathit{power}(m,n) = \begin{cases} 0, & \text{if } m = 0; \\ \mathit{sgn}(\mu k \leqslant m[n^k = m]), & \text{otherwise.} \end{cases}$$

(d) $$prime(n) = \begin{cases} 0, & \text{if } n \leqslant 1; \\ 1, & \text{if } n = 2; \\ \overline{sgn}\left(\mu k \leqslant n\left[(1 < k < n) \wedge rm(n,k) = 0\right]\right), & \text{otherwise.} \end{cases}$$

(e) $$pr(0) = 2;$$
$$pr(n+1) = \mu z \leqslant ((pr(n))! + 1)\left[prime(z) \wedge pr(n) < z\right].$$

In (e) the bound has been constructed according to the following reasoning. If k is prime, then there must exist at least 1 prime between k and $k! + 1$, since no prime less than or equal to k can divide $k! + 1$, and every $j \geqslant 2$ is divisible by some prime. □

Often in elementary mathematics the recursive step in a definition of a function involves a number of earlier values of that function. For example, the Fibonacci sequence 1, 1, 2, 3, 5, 8, . . . is defined by

$$a_0 = 1,$$
$$a_1 = 1,$$
$$a_{n+1} = a_n + a_{n-1}.$$

Is this sequence primitive recursive? Using the pairing and decoding functions τ, π_1, and π_2 of Chapter 3 it is easy to show that this is so. (Why are τ, π_1, and π_2 primitive recursive?) Define

$$b_0 = \tau(1,1),$$
$$b_1 = \tau(2,1),$$
$$b_{n+1} = \tau(a_n, a_{n-1}) = \tau\left(plus(\pi_1(b_n), \pi_2(b_n)), \pi_1(b_n)\right).$$

The function $b(n) = b_n$ is clearly primitive recursive, and $\pi_1 \circ b$ generates the Fibonacci sequence. In our next result we show that any definition which mentions earlier arguments in the recursive step in a reasonable way is an admissible primitive recursive statement.

11 Definition. Let $f(n)$ be a primitive recursive function. Then the *history function* for f, f^*, is defined as follows:

$$f^*(0) = 1,$$
$$f^*(n+1) = \prod_{i=0}^{n} (pr(i))^{f(i)+1}.$$

If $fact(n)$ is the factorial function, then $fact^*(3) = 2^{0!+1} \cdot 3^{1!+1} \cdot 5^{2!+1}$.

The prime power encoding used in the history function is useful because it gives us primitive recursive encoding for finite sequence. It also comes equipped with primitive recursive decoding functions d_k, $k \in \mathbf{N}$:

$$d_k(n) = \mu j \leqslant n\left[rm\left(n, pr(k)^{j+1}\right) = 0 \wedge rm\left(n, pr(k)^{j+2}\right) \neq 0\right].$$

So

$$d_k(f^*(n)) = \begin{cases} f(k), & \text{if } k < n; \\ 0, & \text{otherwise.} \end{cases}$$

Since prime power expressions are unwieldy, we will sometimes write $f^*(n+1)$ as $\langle f(0), f(1), \ldots, f(n)\rangle$. In this notation $fact^*(3) = \langle 0!, 1!, 2!\rangle = \langle 1, 1, 2\rangle$.

12 Theorem (Course of Values Recursion). *Let g be a primitive recursive function of one variable. Then the function f defined by*

$$f(n) = g(f^*(n))$$

where f^* is the history function for f, is also primitive recursive.

PROOF. Consider f^*. Below we tabulate several of its values.

$$f^*(0) = 1,$$

$$f^*(1) = \langle f(0)\rangle = \langle g(1)\rangle,$$

$$f^*(2) = \langle f(0), f(1)\rangle = \langle g(1), g(f^*(1))\rangle = \langle g(1), g(\langle g(1)\rangle)\rangle,$$

$$f^*(3) = \langle g(1), g(\langle g(1)\rangle), g(\langle g(1), g(\langle g(1)\rangle)\rangle)\rangle.$$

Now f^* is primitive recursive *independent* of f:

$$\begin{aligned} f^*(0) &= 1, \\ f^*(n+1) &= f^*(n) \cdot pr(n)^{g(f^*(n))+1} \\ &= \langle \underbrace{g(1), g(\langle g(1)\rangle), \ldots,}_{\alpha} g(\langle \alpha\rangle)\rangle. \end{aligned}$$

This permits us to use f^* to define f:

$$\begin{aligned} f(0) &= g(1), \\ f(n+1) &= d_{n+1}(f^*(n+2)). \end{aligned}$$ □

As with most of our other results, Theorem 12 extends easily to functions involving parameters (Exercise 9).

Course of values recursion allows us to handle another important special form of primitive recursion, *simultaneous recursion*.

13 Definition. Suppose f_0, g_0, f_1, and g_1 are primitive recursive. We say functions f and g are defined from f_0, g_0, f_1, and g_1 by *simultaneous* *recursion* if f and g are defined by the equations

$$\begin{aligned} f(0,x) &= f_0(x), \\ g(0,x) &= g_0(x), \\ f(n+1,x) &= f_1(f(n,x),\ g(n,x),n,x), \\ g(n+1,x) &= g_1(f(n,x),\ g(n,x),n,x) \end{aligned}$$

That is, f and g are defined in such a way that their values depend on earlier values of both f and g.

14 Theorem. *Suppose f and g are defined by simultaneous recursion from f_0, g_0, f_1, and g_1, as in Definition* 13. *Then f and g are primitive recursive.*

PROOF. Define

$$h(n,x) = \begin{cases} f(n/2,x), & \text{if } n \text{ is even;} \\ g(n-1/2,x), & \text{if } n \text{ is odd.} \end{cases}$$

Since

$$\begin{aligned} f(n,x) &= h(2n,x), \\ g(n,x) &= h(2n+1,x), \end{aligned}$$

f and g are primitive recursive if h is, by the composition rule. Appealing to course of values recursion we have

$$\begin{aligned} h(0,x) &= f_0(x), \\ h(1,x) &= g_0(x), \end{aligned}$$

$$h(n+1,x) = \begin{cases} f_1(d_n(h^*(n+1,x)), d_{n-1}(h^*(n+1,x)), n, x), \\ \qquad\qquad\qquad\qquad \text{if } n \text{ is even;} \\ g_1(d_{n-1}(h^*(n+1,x)), d_n(h^*(n+1,x)), n, x), \\ \qquad\qquad\qquad\qquad \text{if } n \text{ is odd.} \end{cases}$$

□

Theorem 14 generalizes in an obvious way to simultaneous recursion definitions involving j rather than 1 parameters and k rather than 2 functions.

We conclude our treatment of the primitive recursive functions by characterizing them as the functions computable by a restricted form of **while**-programs called **loop**-programs. This characterization will help us establish that the partial recursive functions are an acceptable programming system.

loop-PROGRAMS AND loop FUNCTIONS

A **loop**-*program* is a **while**-program with a restriction placed on its **while** statements. In a **loop**-program every **while** statement must appear in the following context:

```
        ⋮
        Y := 0;
        while X ≠ Y do
            begin 𝕊;
                  Y := succ(Y)
            end
        ⋮
```

where $\mathbb{S}$ is an arbitrary statement *which does not mention variables X and Y*. It is clear that a restricted **while** statement of this kind—a **loop** statement—means "repeat the action specified by statement $\mathbb{S}$ exactly X times". The **loop** statement macro **loop** X **do** $\mathbb{S}$ expands to the **while**-program written above, and is represented diagrammatically as

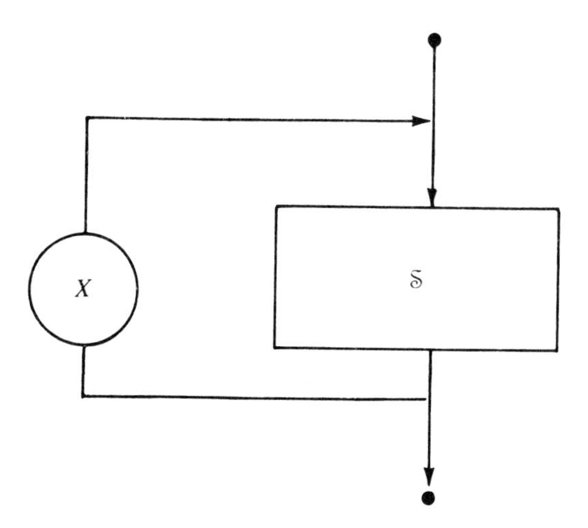

Notice that any loop-free **while**-program is automatically a **loop**-program, as the case with simple assignment, $X := Y$. The next two examples give the **loop** versions of *plus* and *times*.

15 Example. The following **loop**-program adds X and Y and leaves the resulting value in Z:

```
begin
    Z := 0;
    loop X do Z := succ(Z);
    loop Y do Z := succ(Z)
end
```

This **loop** program is represented by the following diagram:

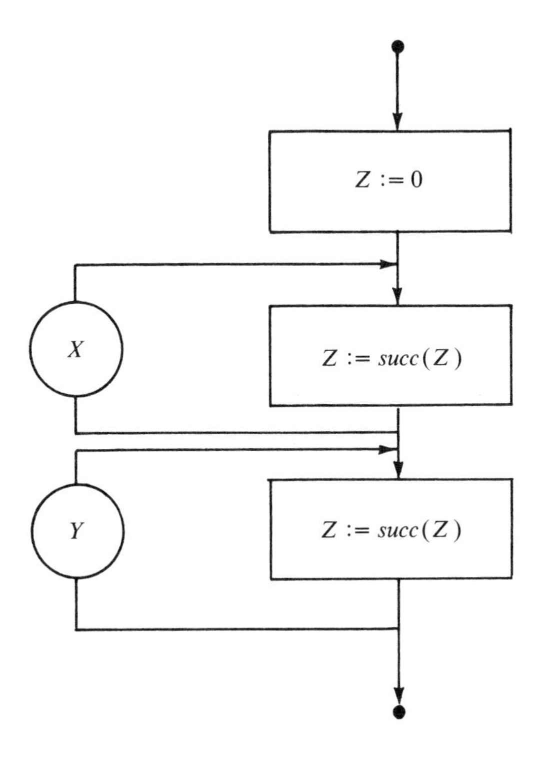

16 Example. The following **loop**-program multiplies the contents of X and Y, and leaves the result in Z.

```
begin
    Z := 0;
        loop X do
            loop Y do
                Z := succ(Z)
end
```

Diagramatically this is shown atop page 216.

Now compare the structure of the programs in Examples 15 and 16. Multiplication requires *loop nesting* of depth 2, while addition requires only nesting of depth 1. It turns out—although we will not prove it here—that there is a direct, computational relationship between the structure complexity of a **loop**-program, measured by the depth of loop nesting, and the running time of the program.[1] Such a relationship is conspicuously absent for the full language of **while**-programs. Indeed, as we have seen, the

[1] This relationship is examined in Meyer and Ritchie (1967). A more accessible reference is Brainerd and Landweber (1974), Chapter 10.

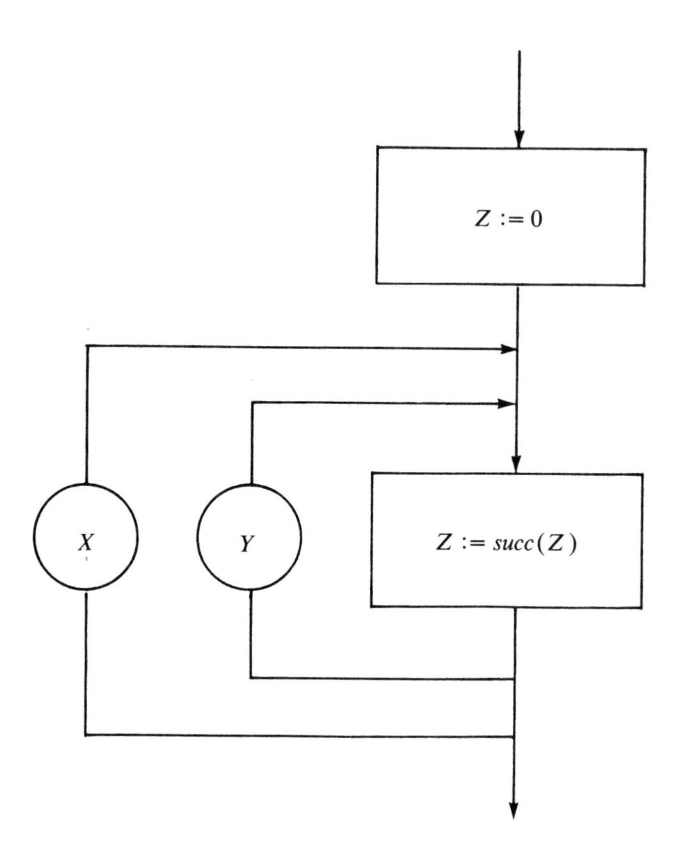

syntactic form of an arbitrary **while**-program in general bears *no* relationship to the ultimate behavior of the program on an arbitrary argument.

Since we are treating the **loop** construction as a **while**-program macro, **loop**-programs inherit **while**-program semantics. Hence we have:

17 Definition. A function $f: \mathbf{N}^j \to \mathbf{N}$ is a **loop** function if it can be computed by some **loop**-program.

Next we give several more detailed examples of **loop** functions. Of course we show later that the **loop** functions coincide with the primitive recursive functions, so the richness of the class of **loop** functions should come as no surprise.

18 Proposition. *Let $\mathcal{S}_1$, $\mathcal{S}_2$, and $\mathcal{S}$ be loop-programs, and let $\mathcal{C}$ be a test (as defined in Section 2.2). Then the following constructions are also* **loop**-*programs*:

(a) **if** $\mathcal{C}$ **then** $\mathcal{S}$
(b) **if** $\mathcal{C}$ **then** $\mathcal{S}_1$ **else** $\mathcal{S}_2$

SKETCH OF PROOF. As in Proposition 2 of Section 2.2, we write an expression $E_{\mathcal{C}}$ involving only the variables appearing in $\mathcal{C}$ such that

$$E_{\mathcal{C}} = \begin{cases} 1, & \text{if } \mathcal{C} \text{ is } \textit{true}; \\ 0, & \text{if } \mathcal{C} \text{ is } \textit{false}. \end{cases}$$

$E_{\mathcal{C}}$ is written in terms of functions computable by **loop**-programs.

Assume X is a new variable not occuring in $\mathcal{C}$, $\mathcal{S}$, $\mathcal{S}_1$, and $\mathcal{S}_2$. The desired **loop**-programs are:

(a) **begin** $X := E_{\mathcal{C}}$;
  loop X **do** $\mathcal{S}$
 end

(b) **begin** $X := E_{\mathcal{C}}$;
  if $X = 1$ **then** $\mathcal{S}_1$;
  if $X = 0$ **then** $\mathcal{S}_2$
 end □

Many (and in fact nearly all!) common number-theoretic and string-processing functions are **loop** functions. The next proposition lists some of these.

19 Proposition. *The following functions are* **loop** *functions.*

(a) x^y,
(b) $x \dot{-} y$,
(c) $x < y$,
(d) $x = y$,
(e) $head(x)$,
(f) $tail(x)$,
(g) $x \| y$,
(h) $\tau(x, y)$,
(i) π_1 *and* π_2,
(j) 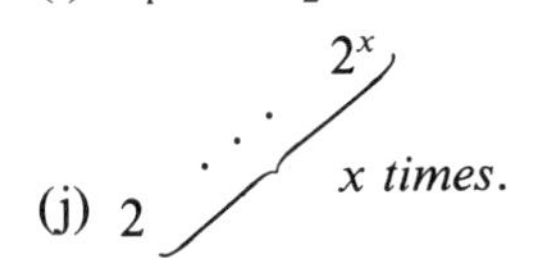 $2^{2^{\cdot^{\cdot^{\cdot^{2^x}}}}}$ *x times.*

PROOF. Exercise 11. □

Our next result shows that the **loop** functions have the same limitation as the primitive recursive functions: they are all total.

20 Proposition. *All* **loop** *functions are total.*

PROOF. By induction on the depth of nesting of loops in **loop**-programs. If nesting depth is 0, the program is made up only of assignment statements, and therefore computes a total function. If the result holds for all **loop**-

programs with loop nesting $\leqslant n$, then consider a program with loop nesting $n+1$. Its key segment must look like

$$\textbf{loop } X \textbf{ do } \mathbb{S}$$

where $\mathbb{S}$ has loop nesting at most n. By the inductive hypothesis, $\mathbb{S}$ is total, and hence application of $\mathbb{S}$ some X times must always return a result, since $\mathbb{S}$ may not alter X. □

We now prove our main result about **loop** functions.

21 Theorem. $f: \mathbf{N}^j \to \mathbf{N}$ *is primitive recursive if and only if it is a* **loop** *function.*

PROOF. Suppose f is primitive recursive. If f is an initial function, then it is certainly a **loop** function. If f is defined by the composition rule, say

$$f(x_1, \ldots, x_j) = h(g_1(x_1, \ldots, x_j), \ldots, g_k(x_1, \ldots, x_j)),$$

where $h, g_1, \ldots, g_k$ are **loop** functions by the inductive hypothesis, then we can compute f with the **loop**-program

```
begin
    Y1 := g1(X1, ..., Xj);
      :
    Yk := gk(X1, ..., Xj);
    X1 := h(Y1, ..., Yk)
end
```

Suppose, therefore, that f is defined by primitive recursion from g and h, where by induction g and h are **loop** functions. We have

$$f(0, x_2, \ldots, x_j) = g(x_2, \ldots, x_j),$$
$$f(n+1, x_2, \ldots, x_j) = h(f(n, x_2, \ldots, x_j), n, x_2, \ldots, x_j).$$

Then the following **loop**-program computes f:

```
begin
    Y := g(X2, ..., Xj); Z := 0;
    loop X1 do
        begin Y := h(Y, Z, X2, ..., Xj); Z := succ(Z) end
    X1 := Y
end
```

Thus every primitive recursive function is a **loop** function.

We prove that every **loop** function is primitive recursive in a slightly stronger form. We show by induction on the depth of nesting of loops in a **loop**-program with k variables that the k functions with values determined by the contents of variable Xi, $1 \leqslant i \leqslant k$, after program execution are all primitive recursive. (We are only required to show that the function determined by $X1$ is primitive recursive.)

The basis step is again trivial: For loop-free programs, i.e., programs with depth of nesting 0, the contents of any variable Xi after execution are determined by finite compositions of *succ*, *pred*, and *zero* assignment, and are therefore clearly primitive recursive.

Now consider **loop** Z **do** $\mathbb{S}$, where $\mathbb{S}$ has depth of nesting of loops $\leqslant n$. We can represent the state of affairs informally by means of the following diagram,

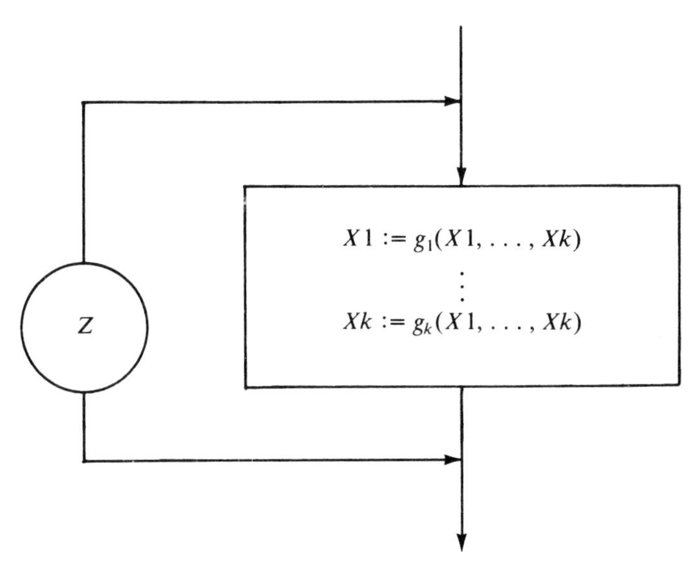

where the functions $g_1, \ldots, g_k$ represent the maps to each variable determined by program $\mathbb{S}$. Since $\mathbb{S}$ has depth of loop nesting $\leqslant n$, by the inductive hypothesis, each g_i, $1 \leqslant i \leqslant k$, is primitive recursive. We can therefore define maps $f_1, \ldots, f_k$ which determine the k variable values of **loop** Z **do** $\mathbb{S}$ using the principle of simultaneous recursion (Theorem 14) as follows:

$$
\begin{aligned}
f_1(0, x_1, \ldots, x_k) &= x_1, \\
&\vdots \\
f_k(0, x_1, \ldots, x_k) &= x_k, \\
f_1(z+1, x_1, \ldots, x_k) &= g_1(f_1(z, x_1, \ldots, x_k), \ldots, f_k(z, x_1, \ldots, x_k)), \\
&\vdots \\
f_k(z+1, x_1, \ldots, x_k) &= g_k(f_1(z, x_1, \ldots, x_k), \ldots, f_k(z, x_1, \ldots, x_k)).
\end{aligned}
$$

Hence every **loop** function is primitive recursive. □

Theorem 21 will be the key result linking **while**-program computations with Kleene's partial recursive functions. We exploit it in our next theorem, which permits us to associate the universal program of Chapter 3 with a particular primitive recursive function. Recall from Section 3.2 that program *short*(e) is $\mathbf{P}_e$ with a limited number of variables.

22 Theorem. *For every* $j > 0$ *there is a* $(j+3)$*-ary* **loop** *function* $t^{(j)}(e, x_1, \ldots, x_j, y, z)$ *such that*

$$t^{(j)}(e, x_1, \ldots, x_j, y, z) = \begin{cases} 1, & \text{if } \mathbf{P}_{short(e)} \text{ on input } x_1, \ldots, x_j \text{ halts in} \leqslant y \\ & \text{steps with output } z; \\ 0, & \text{otherwise.} \end{cases}$$

Sketch of Proof. Consider the interpreter as presented on page 59. It is not hard to show (Exercise 12) that *legal*, *short*, *quad*, *max*, and *next* are all **loop** functions. Indeed the segment:

```
while LBL ≠ MAX do
    begin
      ⋮
    end
```

is the only non-**loop** in the program. Moreover since each **while-do** iteration performs a single instruction of *short*(e), if we suitably bound loop execution we will have our t predicate. Accordingly we convert the interpreter program to the following form:

```
begin if legal(E) ≠ 1 then X1 := 0;
      E := short(E);
      LIST = quad(E);
      LBL := '1';
      MAX := max(LIST);
      loop y do
          begin
            ⋮
          end;
      if LBL = MAX and X1 = z then X1 := 1 else X1 := 0
end
```

This program computes $t^{(j)}(e, x_1, \ldots, x_j, y, z)$. □

Theorem 22 is quite a remarkable result. It says that a (truncated) interpreter which can perform a bounded simulation of any **while**-program is actually a **loop**-program. And even more is true. It is possible to show that t is computable by a **loop**-program with depth of nesting only 2 [see Meyer and Ritchie (1967)].[2]

We now turn our attention to the partial recursive functions, Kleene's formulation of computability, and show that this class, when suitably restricted, is an acceptable programming system.

[2] Our function t is essentially Kleene's T-predicate [Kleene (1952), p. 330].

THE PARTIAL RECURSIVE FUNCTIONS

In order to extend the primitive recursive functions to a class including all computable functions, we permit the restricted application of the following operator.

23 Definition. Let $p(x, y_1, y_2, \ldots, y_n)$ be any $(n+1)$-ary function. The *minimization operator* μx applied to $p(x, y_1, \ldots, y_n)$, which we write as

$$\mu x\big[p(x, y_1, \ldots, y_n) = 1\big],$$

is defined to be the least x such that $p(x, y_1, \ldots, y_n) = 1$, if such an x exists, and is undefined otherwise.

24 Proposition. *Let $p(x, y_1, \ldots, y_n)$ be any total computable function. Then $f(y_1, \ldots, y_n) = \mu x[p(x, y_1, \ldots, y_n) = 1]$ is computable.*

PROOF. Exercise 13. □

The class of partial recursive functions, Kleene's formal characterization of computability, are defined inductively as follows.

25 Definition (The Partial Recursive Functions). A function $\psi : \mathbf{N}^k \to \mathbf{N}$ is partial recursive if

(a) it is one of the primitive recursive basis functions;
(b) it is constructed by composition of partial recursive functions;
(c) it is constructed by primitive recursion from partial recursive functions;
(d) it is constructed by application of the μ-operator to a total partial recursive function, i.e., by an expression of the form

$$f(x) = \mu y\big[p(x, y) = 1\big],$$

where p is a total partial recursive function.

Notice that condition (d) is somewhat peculiar: the μ-operator may be applied only to *total* partial recursive functions, which form a nonrecursive class. But as we shall shortly see, it is sufficient to restrict partial recursive definitions to one application of the μ-operator, applied to a single primitive recursive predicate. This "normal form" result will allow us to enumerate the partial recursive functions. We begin by proving that the partial recursive functions are complete.

26 Theorem. *A function ψ is partial recursive if and only if it is* **while**-*program computable.*

PROOF. If $\psi : \mathbf{N}^k \to \mathbf{N}$ is partial recursive, then by induction on the structure of ψ's definition, ψ is **while**-program computable. If ψ is the *successor* function, the zero function, or a projection function, then ψ is obviously **while**-program computable. Similarly, the composition of **while**-program computable functions is **while**-program computable. Suppose that ψ is defined from ρ and σ by primitive recursion:

$$\psi(0, x_1, \ldots, x_j) = \sigma(x_1, \ldots, x_j),$$

$$\psi(n+1, x_1, \ldots, x_j) = \rho(\psi(n, x_1, \ldots, x_j), n, x_1, \ldots, x_j),$$

where by induction hypothesis, σ and ρ are **while**-program computable. Then the following program computes ψ, leaving the final result in the variable Y:

```
begin
    Y := σ(X1, . . . , Xj);
    Z := n;
    while Z ≠ 0 do
        begin
            Y := ρ(Y, n − Z, X1, . . . , Xj);
            Z := pred(Z)
        end
end
```

Finally, if $\psi(x_1, \ldots, x_j) = \mu y[p(y, x_1, \ldots, x_j) = 1]$ where p is total and **while**-program computable, then the following program will compute ψ, leaving the result in variable Y:

```
begin
    Y := 0;
    Z := p(0, X1, . . . , Xj);
    if Z ≠ 1 then
    while Z ≠ 1 do
        begin
            Y := succ(Y);
            Z := p(Y, X1, . . . , Xj)
        end
end
```

Suppose next that $\varphi_e^{(j)}$ is an arbitrary **while**-program computable function. Then by Theorem 22, we have

$$\varphi_e^{(j)}(x_1, \ldots, x_j) = \pi_2\Big(\mu y\Big[t^{(j)}(e, x_1, \ldots, x_j, \pi_1(y), \pi_2(y)) = 1\Big]\Big). \quad (\$)$$

Since $t^{(j)}$ is a **loop** function, by Theorem 21 it is also a primitive recursive function. But then $\varphi_e^{(j)}$ is partial recursive, since π_1, π_2, and t are all primitive recursive. □

The assertion ($)—that any computable function can be represented as a decoding of a μ-application applied to a fixed primitive recursive function —is known as *Kleene's Normal Form Theorem*. It is a striking result, asserting that a single unbounded search applied to a predicate of very low complexity is sufficient to represent any computable function. It also gives us a way to enumerate the partial recursive functions. Let $\psi_0, \psi_1, \ldots, \psi_e, \ldots$ be the listing of partial recursive functions of one variable, where ψ_e is identified with $\pi_2(\mu y(t(e, x, \pi_1(y), \pi_2(y)) = 1))$. Conveniently, this says that $\varphi_e = \psi_e$ for all e.

We conclude our discussion of the Kleene characterization by establishing that the partial recursive functions as indexed above form an acceptable programming system.

27 Theorem. *The partial recursive functions are an acceptable programming system. That is, they satisfy the s-m-n property and the universal function property.*

PROOF. Using the numbering scheme described above, it turns out that the s-m-n functions are also the parametrization functions for the partial recursive functions, as we illustrate below for the case of the partial recursive functions of two variables:

$$\psi_e^{(2)}(x, z) = \varphi_e^{(2)}(x, z) = \varphi_{s_1^1(e,x)}(z) = \pi_2\Big(\mu y\Big[t\big(s_1^1(e, x), z, \pi_1(y), \pi_2(y)\big) = 1\Big]\Big)$$
$$= \psi_{s_1^1(e,x)}(z).$$

The universal function for the $\{\psi_i^{(j)} \mid i \in \mathbf{N}\}$ is just

$$\Psi(e, x_1, \ldots, x_j) = \pi_2\Big(\mu y\Big[t^{(j)}(e, x_1, \ldots, x_j, \pi_1(y), \pi_2(y)) = 1\Big]\Big). \qquad \square$$

For a proof of Theorem 27 based on a different Gödel numbering of partial recursive functions, see Exercise 14.

EXERCISES FOR SECTION 9.2

1. Give a completely formal primitive recursive definition for *times*, using only Definition 1 and the fact that *plus* is primitive recursive.
2. Complete the proof of Proposition 2.
3. Prove that all polynomials are primitive recursive.
4. Prove that the pairing and decoding functions τ, π_1, and π_2 are primitive recursive.
5. Give a Gödel numbering for the set of primitive recursive functions.
6. Prove Corollary 7.

7. Is the Fermat function given on page 2 primitive recursive?

8. Let $f: \mathbf{N} \to \mathbf{N}$ be a total function of one variable. Define

$$\hat{f}(0,x) = x,$$
$$\hat{f}(n,x) = \underbrace{f \circ f \circ \cdots \circ f}_{n \text{ times}}(x).$$

That is, $\hat{f}(n,x)$ is the value of f composed with itself n times and then applied to x. Prove that if f is primitive recursive, then $\hat{f}$ is primitive recursive.

9. State and prove the course-of-values theorem, Theorem 12, for a function f with parameters.

10. Let B be a set whose characteristic function is primitive recursive, and consider the set

$$A = \{n \mid \mathrm{DOM}(\varphi_n) = B\}.$$

Is A recursive? Is A r.e.?

11. Prove Proposition 19.

12. Show that the functions *legal*, *short*, *quad*, *max*, and *next* in the proof of Theorem 22 are **loop** functions.

13. Prove Proposition 24.

14. Given a Gödel numbering for the primitive recursive functions, $p_0, p_1, \ldots, p_n, \ldots,$ Gödel-number the function class obtained by

   (a) applying the μ-operator no more than once to the p_i's; and
   (b) closing the class off under composition and primitive recursion. Let $\theta_0, \theta_1, \ldots, \theta_n, \ldots$ be a listing of the resulting class. Show that $\{\theta_i \mid i \in \mathbf{N}\}$ is an acceptable programming system.

15. Let g be any total computable function of one variable. Show that there is a two-variable total computable h such that $g(x) = \mu y[h(x, y) = 1]$.

16. Show that the computable functions are not closed under μ. That is, show that there is some (necessarily partial) computable $\theta(x, y)$ such that $\theta(x) = \mu y[\theta(x, y) = 1]$ is not computable.

9.3 Symbol-Manipulation Systems and Formal Languages

In this section we consider three types of symbol manipulation systems, semi-Thue systems, Post systems, and context-free grammars. Of these three, context-free grammars are by far the most important for computer science because they are reasonably adequate for specifying the syntax of programming languages. Indeed we gave a context-free description of the syntax of **while**-programs in Section 2.2.

We shall see that all three of these systems have essential undecidable properties. We show first that a simple "word problem" for semi-Thue systems encompasses the Halting Problem for Turing machines, and is therefore undecidable. This will imply that the Post Correspondence Problem—Does a certain arbitrarily defined set of equations involving strings have a solution?—is also undecidable. Finally we show how the undecidability of the Post Correspondence Problem implies that the problem of determining if an arbitrary context-free grammar is *ambiguous*, i.e., can generate multiple distinct derivations of some string, is undecidable.

In our treatment of symbol-manipulation systems, we will mostly be concerned with finite strings built up from a finite alphabet of symbols. The empty string will be denoted by λ. If Σ is a finite alphabet, we will use the expression Σ^* to denote all finite strings from Σ. Σ^+ denotes all nonempty finite strings.

Semi-Thue systems

Semi-Thue systems, one of the earliest grammatical systems to be studied, were first considered by the Norwegian mathematician Axel Thue early in this century. (Thue is pronounced "two-way".)

1 Definition. *A semi-Thue system* T over a finite alphabet Σ is a pair $T = (\Sigma, P)$ where P is a set of production rules of the form $\alpha \to \beta$, where $\alpha \in \Sigma^+$ and $\beta \in \Sigma^*$. We write $x \Rightarrow_T y, x, y \in \Sigma^*$, to indicate that x can be written as $u\alpha v$, $\alpha \to \beta$ is a production of P, and $u\beta v = y$. We write $x \overset{*}{\Rightarrow}_T y$ if $x = y$ or there is a sequence $w_1, \ldots, w_n$ of strings in Σ^* such that $w_1 = x$, $w_n = y$, and for $1 \leqslant j < n$, $w_j \Rightarrow_T w_{j+1}$. We write $\Rightarrow$ and $\overset{*}{\Rightarrow}$ instead of $\Rightarrow_T$ and $\overset{*}{\Rightarrow}_T$ when the context is clear. If $x \Rightarrow_T y$ we shall say that y is *directly derivable from* x *in* T. If $x \overset{*}{\Rightarrow}_T y$ we shall say that y is *derivable from* x *in* T.

2 Definition. *The word problem for semi-Thue* systems is the problem of determining, for arbitrary strings x and y over Σ and arbitrary semi-Thue system T, whether or not $x \overset{*}{\Rightarrow}_T y$.

3 Example. Given semi-Thue system $T = (\{a, b\}, \{ab \to bbb, bb \to \lambda\})$, we have

(1) $abbab \overset{*}{\Rightarrow} bbbbbbb$,

(2) $babb \overset{*}{\Rightarrow} b$,

(3) $abbabbb \overset{*}{\not\Rightarrow} bbbb$.

4 Theorem. *The word-problem for semi-Thue systems is undecidable.*

PROOF. By Theorem 5 of Section 9.1 we know that the problem of determining if $\tau_e(x) = y$ for arbitrary Turing computable function τ_e is undecidable. Our proof will establish the word-problem undecidability by showing that

$$\tau_e(x) = y \text{ iff } \$q_0B\bar{x}¢ \overset{*}{\Rightarrow}_T \bar{y}$$

where x and y are unary encodings of x and y, \$ and ¢ are special marker symbols, and T is constructed from the state structure of Turing machine e with start state q_0.

Our proof will proceed with a particular example in mind, the Turing machine presented below:[3]

$$\begin{array}{l}
(q_0 \quad B \quad q_1 \quad B \quad R)\\
(q_1 \quad 1 \quad q_1 \quad 1 \quad R)\\
(q_1 \quad B \quad q_2 \quad 1 \quad L)\\
(q_2 \quad 1 \quad q_2 \quad 1 \quad L)\\
(q_0 \quad 1 \quad q_H \quad 1 \quad N)\\
(q_2 \quad B \quad q_H \quad B \quad N)
\end{array}$$

Since our translation method from Turing machines to semi-Thue systems will be completely general, our demonstration by example will have the force of proof.

There are two stages to our translation method. First, T simulates machine execution. Then, if and when a legal final configuration involving q_H appears, T "cleans up" the string by removing B's and marker symbols, leaving the computed integer coded in unary as "output". T is defined over the alphabet $\Sigma = \{\$, ¢, 1, A, B, C, D, E, q_0, \ldots, q_H\}$.

Stage 1. Simulation begins on the string \$ q_0 B $\bar{x}$ ¢. That is, the initial string has a left marker \$, a right marker ¢, a current state q_0, and a unary encoding of integer x.

Then T "executes" the transition rules of the machine. Thus the quintuple

$$(q_0 \quad B \quad q_1 \quad B \quad R)$$

becomes the rewrite rule

$$q_0 \quad B \to B \quad q_1.$$

The quintuple

$$(q_2 \quad 1 \quad q_2 \quad 1 \quad L)$$

[3] This is the Turing machine of Example 1 of Section 9.1.

leads to two rewrite rules:

$$1 \quad q_2 \quad 1 \rightarrow q_2 \quad 1 \quad 1$$
$$B \quad q_2 \quad 1 \rightarrow q_2 \quad B \quad 1.$$

We also need rewrite rules that will allow us to construct a potentially infinite tape. We accomplish this by adding blank symbols at both ends of the string:

$$\$ \rightarrow \$ \quad B$$
$$¢ \rightarrow B \quad ¢.$$

Stage 2. Our Turing machine semantics conventions require that upon termination the actual machine tape looks like

$$\ldots B \quad \underset{q_H}{\overset{\blacktriangle}{B}} \quad 1 \quad 1 \quad 1 \ldots 1 \quad 1 \quad B \ldots$$

corresponding to the derived string

$$\ldots B \quad q_H \quad B \quad 1 \quad 1 \quad 1 \ldots 1 \quad 1 \quad B \ldots$$

Hence we design Stage 2 to "clean up" this configuration, yielding as a final string the first output block of 1's on the tape. Here are the rules:

$$q_H B \rightarrow A$$
$$A1 \rightarrow 1A$$
$$AB \rightarrow C$$
$$CB \rightarrow C$$
$$C1 \rightarrow C$$
$$C¢ \rightarrow D$$
$$1D \rightarrow D1$$
$$D \rightarrow E$$
$$BE \rightarrow E$$
$$\$E \rightarrow \lambda$$

The effect of the rules is to (1) exchange q_H for A, and pass A through the first block of 1's; (2) switch control to a new "state", the symbol C, which eliminates all symbols to the right of the first output block; (3) pass new symbol D through the first block right to left; and (4) eliminate all B's and the \$ to the left of the first block. Notice that if 1's occur to the left of the head, then E will never "reach" \$ in the semi-Thue system, and the resulting string will be invalid.

Thus $\$q_0 B\bar{x}¢ \overset{*}{\Rightarrow}_T \bar{y}$ if and only if the machine halts on input x with output y. Our translation technique from machines to semi-Thue systems is completely general, and the result follows. □

The connection between semi-Thue systems and computable functions is examined further in Exercise 9.

Next we consider a different kind of undecidability result for string processing by Post systems. Post introduced his equation-based systems in 1946 [Post (1946)].

5 Definition. A *Post system P* over a finite alphabet Σ is a finite set of ordered pairs $\{(\alpha_i, \beta_i)\}$, $1 \leqslant i \leqslant n$, where $\alpha_i, \beta_i \in \Sigma^*$. The *Post Correspondence Problem* (PCP) is the problem of determining whether, for an arbitrary Post system P, there exist integers $i_1, \ldots, i_k$ such that

$$\alpha_{i_1} \ldots \alpha_{i_k} = \beta_{i_1} \ldots \beta_{i_k}.$$

The i_j's need not be distinct.

6 Example. Consider the Post system over $\Sigma = \{0, 1\}$:

$$P = \{(1, 01), \quad (\lambda, 11), \quad (0101, 1), \quad (111, \lambda)\}.$$

This system has a solution since

$$\alpha_2\alpha_1\alpha_1\alpha_3\alpha_2\alpha_4 = \beta_2\beta_1\beta_1\beta_3\beta_2\beta_4.$$

Thus the Post Correspondence Problem asks if a set of equations involving strings has a solution of the special form described above.

7 Theorem. *The Post Correspondence Problem is undecidable.*

PROOF. We reduce the word problem for semi-Thue systems to the PCP. Given an arbitrary semi-Thue system $T = (\Sigma, P)$, we expand the alphabet of Σ to Σ' as follows:

(1) For every $a \in \Sigma$, we place a new symbol $\bar{a} \in \Sigma'$;
(2) we add new symbols $*, \bar{*}, [\ , \ ]$ to Σ'.

If w is a string in Σ^*, we write $\bar{w}$ for the string constructed by replacing every symbol of w by its barred companion.

Given a word problem $x \overset{*}{\Rightarrow}_T y$, we transform it into a Post Correspondence Problem for a Post system which is suitable for encoding semi-Thue replacement sequences of the form $w_1 \Rightarrow w_2 \Rightarrow \cdots \Rightarrow w_n$, with $x = w_1$ and $y = w_n$. The terms of the Post equations for such a sequence will have a special form. First, we force any possible PCP solution to begin with

$$\begin{array}{l} [w_1 * \\ [\end{array}$$

(The top line corresponds to the left-hand side of the equation, and the bottom line to the right-hand side.) Next, we force possible solutions to look like:

$$\begin{array}{l} [w_1 * \bar{w}_2 \bar{*} \\ [w_1 * \end{array}$$

Continuing, we require

$$[w_1 * \overline{w}_2 \bar{*} w_3 *$$
$$[w_1 * \overline{w}_2 \bar{*}$$

and so forth, alternating barred and unbarred blocks as we go, until

$$[w_1 * \overline{w}_2 \bar{*} \ldots \overline{w}_{n-1} \bar{*} w_n$$
$$[w_1 * \overline{w}_2 \bar{*} \ldots \overline{w}_{n-1}$$

is reached. (Later we describe how to force the last term to be unbarred.) Finally, the second line "catches up", and we get on both lines

$$[w_1 * \overline{w}_2 \bar{*} \ldots w_n]$$

Given semi-Thue system T and words x and y, the corresponding Post rules are:

(1) $(a, \bar{a})$ and $(\bar{a}, a)$ for each $a \in \Sigma \cup \{*\}$,
(2) $([w_1 *, [)$, where $w_1 = x$,
(3) $(], \bar{*} w_n])$, where $w_n = y$,
(4) $(\beta, \bar{\alpha})$ and $(\bar{\beta}, \alpha)$ for each rule $\alpha \to \beta$ of T.

Clearly (2) must be the first rule invoked if a match is to be found, since it is the only rule with matching first symbols. Similarly, (3) must be the last rule used.

Suppose now that the intermediate form

$$[w_1 * \ldots \bar{*} w_k * \overline{w}_{k+1} \bar{*}$$
$$[w_1 * \ldots \bar{*} w_k *$$

has been reached, for $k + 1 < n$. Suppose $w_{k+1} \Rightarrow w_{k+2}$ by the rule $\alpha \to \beta$, with $w_{k+1} = u\alpha v$, $w_{k+2} = u\beta v$. To mimic this transition in the Post system, we write

$$[w_1 * \ldots \bar{*} w_k * \overline{w}_{k+1} \bar{*} u$$
$$[w_1 * \ldots \bar{*} w_k * \bar{u}$$

by repeated application of the rule (1). By rule (4) we write

$$[w_1 * \ldots \bar{*} w_k * \overline{w}_{k+1} \bar{*} u\beta$$
$$[w_1 * \ldots \bar{*} w_k * \bar{u}\bar{\alpha}$$

And by applying rule (1) again we get

$$[w_1 * \ldots \bar{*} w_k * \overline{w}_{k+1} * u\beta v$$
$$[w_1 * \ldots \bar{*} w_k * \bar{u}\bar{\alpha}\bar{v}$$

which is just

$$\begin{array}{l}[w_1 * \ldots \bar{*} w_k * \bar{w}_{k+1} * w_{k+2} \\ [w_1 * \ldots \bar{*} w_k * \bar{w}_{k+1}\end{array}$$

One final point. Notice that w_n in rule (3) is not barred. This can apparently happen only if n is odd, since the "parity" of strings changes from block to block. However, using rule (1) repeatedly we can always change a pair of the form

$$\begin{array}{l}[w_1 * \ldots \bar{*} w_k * \bar{w}_{k+1} \bar{*} \\ [w_1 * \ldots \bar{*} w_k *\end{array}$$

into

$$\begin{array}{l}[w_1 * \ldots \bar{*} w_k * \bar{w}_{k+1} \bar{*} w_{k+1} * \\ [w_1 * \ldots \bar{*} w_k * \bar{w}_{k+1} \bar{*}\end{array}$$

thus reversing the parity of the final block.

A simple inductive argument can now be used to show that $x \overset{*}{\Rightarrow}_T y$ if and only if the constructed Post system has a solution, and this completes the reduction. □

Next we consider context-free grammars. [For a detailed elementary introduction to these grammars and the languages they generate, see Arbib, Kfoury, and Moll (1981)]. Context-free grammars resemble semi-Thue systems, except that their production rules are of a restricted form and they operate over two alphabets, a finite terminal symbol alphabet V_T and a finite nonterminal symbol alphabet V_N. The notion of derivability coincides with the corresponding notion for semi-Thue systems: "$\Rightarrow$" means "directly derivable" and "$\overset{*}{\Rightarrow}$" means "derivable".

In our grammatical description of **while**-programs in Section 2.1 we used the symbol ::= to indicate that one string of symbols may be replaced by another, as in ⟨test⟩ ::= ⟨variable⟩ ≠ ⟨variable⟩. Below we use the simpler notation → for string replacement.

8 Example (Matched Parentheses). The set of strings derivable from the symbol S using the following replacement rules coincides with the set of legal matched parentheses:

(1) $S \to (\ )$,
(2) $S \to (S)$,
(3) $S \to SS$.

To derive the string (( )( )) from S apply the following productions:

$$S \underset{(2)}{\Rightarrow} (S) \underset{(3)}{\Rightarrow} (SS) \underset{(1)}{\Rightarrow} ((\ )S) \underset{(1)}{\Rightarrow} ((\ )(\ ))$$

In the grammar of Example 8 there are two terminal symbols, $V_T = \{(\ ,\ )\}$, one nonterminal symbol, $V_N = \{S\}$, the three production rules given above, and a start symbol, S. Notice that every production has a single nonterminal as its left-hand side. This is the distinguishing characteristic of context-free grammars, and it allows us to represent derivations as trees. For instance the derivation of (( )( )) in Example 8 can be represented by the tree

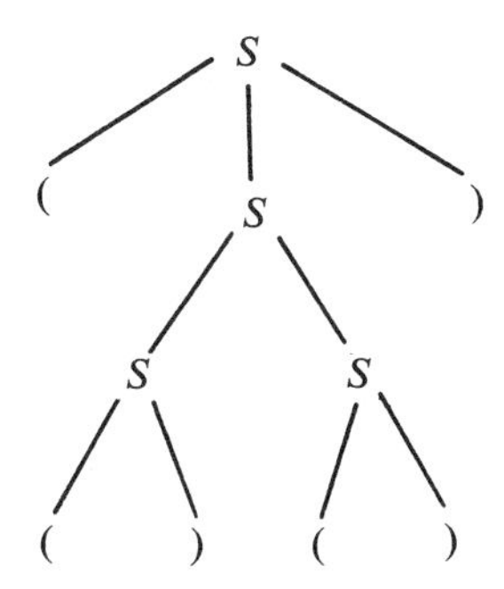

9 Definition. A *context-free grammar* G is a quadruple $G = (V_N, V_T, S, P)$ where V_N is a nonterminal alphabet, V_T is a terminal alphabet, S is the start symbol, and P is the set of productions. Every production of P must be of the form $A \to \alpha$, where $A \in V_N$ and α is a nonempty string of terminal and nonterminals. (The condition that α not be empty can be relaxed. See Exercise 5.)

10 Definition. Define $L(G)$, the *context-free language* generated by context-free grammar G, to be

$$L(G) = \left\{ x \in V_T^* \mid S \overset{*}{\Rightarrow} x \right\}.$$

That is, $L(G)$ is the set of all strings of terminal symbols derivable from the start symbol S.

If G is the grammar of Example 8, then $L(G)$ is the set of all legally matched parentheses.

11 Example. Consider the grammar G with

$$V_N = \{S, C\},$$

$$V_T = \{\textbf{if}, \textbf{then}, \textbf{else}, a, b, p, q\}$$

The start symbol is S, while P contains the productions

$$S \to \textbf{if } C \textbf{ then } S \textbf{ else } S$$
$$S \to \textbf{if } C \textbf{ then } S$$
$$S \to a \mid b$$
$$C \to p \mid q$$

(Note: $A \to \alpha \mid \beta$ is shorthand for: $A \to \alpha$, $A \to \beta$). This grammar generates precisely the **if-then** and **if-then-else** forms in the language of **while**-programs. For example, we can generate **if** p **then if** q **then** a **else** b:

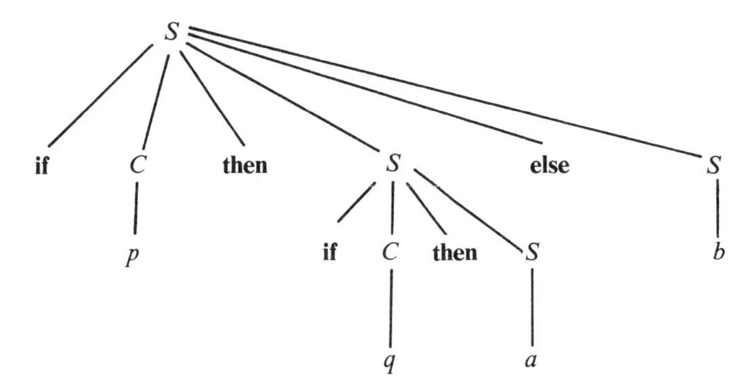

Unfortunately, the grammar of Example 11 is *ambiguous*: certain strings in the language can be derived by more than one derivation tree. For example, **if** p **then if** q **then** a **else** b is derived by the tree on page 233, which is distinct from the one above. It is not hard to give a more careful grammar which is unambiguous, and which conforms to the PASCAL convention of associating each **else** with the nearest **if**.

12 Example. An unambiguous grammar for **if-then-else**:

$$S \to S_1 \mid S_2$$
$$S_1 \to \textbf{if } C \textbf{ then } S_1 \textbf{ else } S_2 \mid a \mid b$$
$$S_2 \to \textbf{if } C \textbf{ then } S \mid \textbf{if } C \textbf{ then } S_1 \textbf{ else } S_2 \mid a \mid b$$
$$C \to p \mid q$$

Context-free grammars are mechanisms for listing strings, and so they bear a resemblance to the machinery we have developed for listing r.e. sets. But unlike r.e. sets, context-free languages have a decidable membership problem.

13 Theorem. *The problem of determining if an arbitrary string x belongs to an arbitrary context-free language $L(G)$ is decidable.*

PROOF. Since context-free production rules never decrease the length of a string, it is possible to list, in order, all derived strings of length 1, length

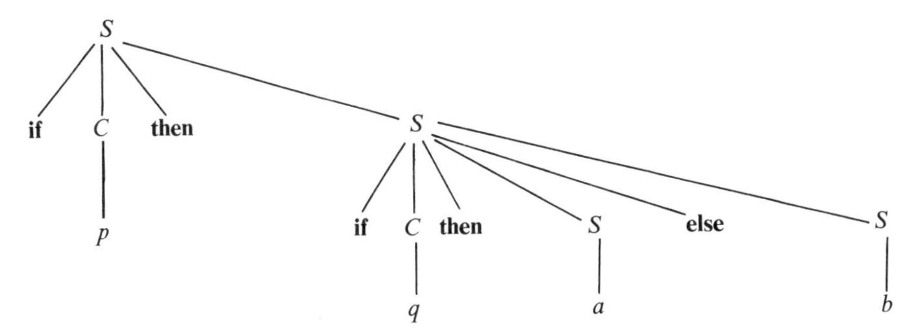

2, ... To determine if $x \in L(G)$, run this procedure until strings of x's length are encountered, and check x against this finite list. □

Of course this is an extremely crude decision procedure. For the special context-free portions of programming languages, language membership can usually be judged by algorithms that run in time proportional to the length of the input string x.

However, a great many properties of context-free grammars *are* undecidable. Below we prove undecidability for the problem of determining grammatical ambiguity.

14 Theorem. *The problem of determining if a context-free grammar G is ambiguous is undecidable.*

Proof. Let $P = \{(\alpha_1, \beta_1), \ldots, (\alpha_n, \beta_n)\}$ be a Post system over Σ. Augment Σ with n new symbols $\bar{1}, \bar{2}, \ldots, \bar{n}$, and consider the following context-free grammar

$$S \to S_1 \mid S_2$$

$$S_1 \to \alpha_1 S_1 \bar{1} \mid \alpha_2 S_1 \bar{2} \mid \cdots \mid \alpha_n S_1 \bar{n} \mid \alpha_1 \bar{1} \mid \cdots \mid \alpha_n \bar{n}$$

$$S_2 \to \beta_1 S_2 \bar{1} \mid \beta_2 S_2 \bar{2} \mid \cdots \mid \beta_n S_2 \bar{n} \mid \beta_1 \bar{1} \mid \cdots \mid \beta_n \bar{n}$$

This grammar is ambiguous if and only if P has a solution. □

We conclude this section by stating, without proof, some other known undecidable properties of context-free grammars and languages.

15 Theorem. *Let G_1 and G_2 be arbitrary context-free grammars. Then the following properties are undecidable.*

(a) *$L(G_1)$ is empty;*
(b) *$L(G_1) \cap L(G_2)$ is empty;*
(c) *$L(G_1) = L(G_2)$;*
(d) *$L(G_1) \subset L(G_2)$.*

Exercises for Section 9.3

1. Is the Post system $P = \{(101, 10), (11111, 11), (\lambda, 1)\}$ solvable?

2. Prove that a Post system over a one symbol alphabet is decidable.

3. Give an effective listing of all Post systems over the alphabet $\Sigma = \{a, b\}$.

4. Let $L = L(G)$ be a context-free language over $V_T = \{0, 1\}$. Define $L^R = \{x^R \mid x \in L\}$, where x^R is the reversal of string x. Prove that L^R is context-free.

5. Let G be a context-free grammar which involves productions $A \rightarrow \lambda$. Construct an equivalent grammar G' with start symbol S' such that:

   (a) $S' \rightarrow \lambda$ is the only production to λ in G';
   (b) S' does not appear on the right-hand side of any production; and
   (c) $L(G) = L(G')$.

6. Construct a grammar that yields the following derivation tree:

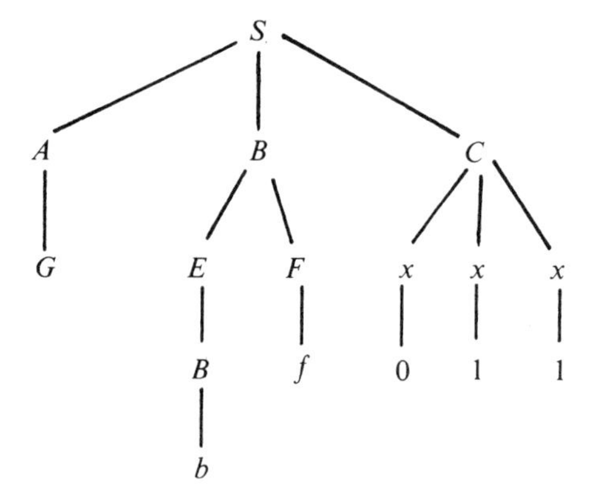

7. What language does the following grammar generate?
$$S \rightarrow SS \mid aSb \mid bSa \mid ab \mid ba$$

8. Write a context-free grammar for the parenthesized language of +, 1, and 2. Example: ((1 + 2) + 2).

9. Let ψ be any computable function. Show that there exists a semi-Thue system $T = (\Sigma, P)$ such that for all m and n, $\psi(m) = n$ if and only if $\overline{m} \overset{*}{\Rightarrow}_T \overline{n}$, where $\overline{m}$ and $\overline{n}$ are m and n in unary notation. State and prove the "universal function property" for semi-Thue systems.

References

Alagic, S., and M. A. Arbib (1978): *The Design of Well-Structured and Correct Programs*, Springer-Verlag.

Arbib, M. A. (1963): "Monogenic Normal Systems are Universal," *J. Australian Math. Soc.*, III, Part 3, 301–306.

Arbib, M. A. (1969): *Theories of Abstract Automata*, Prentice-Hall.

Arbib, M. A., A. J. Kfoury and R. N. Moll (1981): *A Basis for Theoretical Computer Science*, Springer-Verlag.

Arbib, M. A. and E. G. Manes (1982): "The Pattern-of-Calls Expansion is the Canonical Fixpoint for Recursive Definition," *J. ACM, 29*, 577–602.

Brainerd, W. S. and L. H. Landweber (1974): *Theory of Computation*, John Wiley and Sons, New York.

Church, A. (1932): "A Set of Postulates for the Foundations of Logic," *Annals of Mathematics*, *33*, 346–366; and *34*, 839–864.

Church, A. (1936): "An Unsolvable Problem of Elementary Number Theory," *Am. J. Math.*, *58*, 345–363.

Cook, S. A. (1978): "Soundness and Completeness of an Axiom System for Program Verification," *SIAM J. on Computing*, *7*, 70–90.

Davis, M. (1958): *Computability and Unsolvability*, McGraw-Hill Book Company.

Davis, M. (1965): *The Undecidable*, Raven Press, Hewlett, New York.

Davis, M. (1973): "Hilbert's Tenth Problem is Unsolvable," *Am. Math. Monthly*, *80*, 233–269.

Davis, M., H. Putnam and J. Robinson (1961): "The Decision Problem for Exponential Diophantine Equations," *Ann. of Math*, *74*, 425–436.

DeBakker, J. (1980): *Mathematical Theory of Program Correctness*, Prentice-Hall.

Floyd, R. W. (1967): "Assigning Meanings to Programs," Proc. Symp. Applied Math., 19, in J. T. Schwartz (ed.), *Mathematical Aspects of Computer Science*, 19–32, American Mathematical Society, Providence, R.I.

Friedberg, R. M. (1957): "Two Recursively Enumerable Sets of Incomparable Degrees of Unsolvability," *Proceedings of Natl. Academy of Sciences*, *43*, 236–238.

Gödel, K. (1931): "Über Formal Unentscheidbare Sätze der Principia Mathematica und Verwandter Systeme, 1," *Monatschefte für Mathematik und Physik*, *38*, 173–198. (English translation in Davis (1965).)

Hennie, F. (1977): *Introduction to Computability*, Addison-Wesley.
Hoare, C. A. R. (1969): "An Axiomatic Basis for Computer Programming," *Communications of the ACM*, *12*, 576–580, 583.
Hoare, C. A. R. and N. Wirth (1973): "An Axiomatic Definition of the Programming Language PASCAL," *Acta Informatica*, *2*, 335–355.
Jensen, K. and N. Wirth (1974): PASCAL *User Manual and Report*, Lecture Notes in Computer Science, Vol. 18, Springer-Verlag.
Kfoury, A. J. (1974): "Translatability of Schemas over Restricted Interpretations," *J. of Comp. and System Sciences*, *8*, 387–408.
Kleene, S. C. (1935): "A Theory of Positive Integers in Formal Logic," *Amer. J. Math.*, *57*, 153–173; 219–244.
Kleene, S. C. (1936): "General Recursive Functions of Natural Numbers," *Math. Ann.*, *112*, 727–742.
Kleene, S. C. (1952): *Introduction to Metamathematics*, North-Holland, Amsterdam.
Kleene, S. C. (1967): *Mathematical Logic*, Wiley, New York.
Kleene, S. C. (1981): "Origins of Recursive Function Theory," *Annals of the History of Computing*, *3*, no. 1.
Knaster, B. (1928): "Un Theorème sur les Fonctions des Ensembles," *Ann. Soc. Polon. Math.*, *6*, 133–134.
Kozen, D. (1980): "Indexings of Subrecursive Classes," *Theoretical Computer Science*, *11*, 277–301.
Ladner, R. E. (1975): "On the Structure of Polynomial Time Reducibility," *J. ACM*, *22*, 155–171.
Machtey, M. and P. R. Young (1978): *An Introduction to the General Theory of Algorithms*, North-Holland, New York.
Manna, Z. (1974): *Mathematical Theory of Computation*, McGraw-Hill Book Company.
Markov, A. A. (1954): *Theory of Algorithms*, Works of the Steklov Mathematical Institute, Vol. 42 (English translation, 1961, National Science Foundation, Washington, D.C.).
Matijasevic, Y. V. (1970): "Enumerable Sets are Diophantine," *Dokl. Akad. Nauk SSSR*, *191*, 279–282 (in Russian). English translation in *Soviet Math. Dokl.*, *11*, 354–357.
Matijasevic, Y. V. and J. Robinson (1975): "Reduction of an Arbitrary Diophantine Equation to One in 13 Unknowns," *Acta Arith.*, *27*, 521–553.
Meyer, A. R. and D. M. Ritchie (1967): "The Complexity of Loop Programs," *Proc. ACM Natl. Conference*, 465–469.
Minsky, M. L. (1967): *Computation: Finite and Infinite Machines*, Prentice-Hall.
Muchnik, A. A. (1956): "On the Unsolvability of the Problem of Reducibility in the Theory of Algorithms" (Russian), *Doklady Akademii Nauk SSSR*, *108*, 194–197.
Post, E. L. (1936): "Finite Combinatory Processes-Formulation 1," *J. Symbolic Logic*, *1*, 103–105.
Post, E. L. (1943): "Formal Reductions of the General Combinatorial Decision Problem," *American Journal of Math.*, *65*, 197–215.
Post, E. L. (1944): "Recursively Enumerable Sets of Positive Integers and their Decisions Problems," *Bull. Amer. Math. Soc.*, *50*, 284–316.
Post, E. L. (1946): "A Variant of a Recursively Unsolvable Problem," *Bull. Amer. Math. Soc.*, *52*, 264–268.
Post, E. L. (1947): "Recursive Unsolvability of a Problem of Thue," *J. Symbolic Logic*, *12*, 1–11.
Rice, H. G. (1953): "Classes of Recursively Enumerable Sets and Their Decision Problems," *Trans. Amer. Math. Soc.*, *74*, 358–366.

Rice, H. G. (1956): "On Completely Recursively Enumerable Classes and Their Key Arrays," *J. Symbolic Logic*, *21*, 304–308.
Rogers, H. (1958): "Gödel Numberings of Partial Recursive Functions," *J. Symbolic Logic*, *23*, 331–341.
Rogers, H. (1967): *Theory of Recursive Functions and Effective Computability*, McGraw-Hill Book Company.
Scott, D. S. and C. Strachey (1971): "Towards a Mathematical Semantics for Programming Languages," in J. Fox (ed.), *Proc. Symp. on Computers and Automata*, 19–46, Polytechnic Institute of Brooklyn Press, N.Y. (For a textbook account, see Stoy (1977).)
Shepherdson, J. C. and H. E. Sturgis (1963): "Computability of Recursive Functions," *J. ACM*, *10*, 217–255.
Stoy, J. E. (1977): *Denotational Semantics: The Scott-Strachey Approach to Programming Language Theory*, MIT Press, Cambridge, Mass.
Tarski, A. (1936): "Der Wahrheitsbegriff in den Formalisierten Sprachen," *Studia Philosophica*, *1*, 261–405. (English translation in Tarski (1956), 152–278.)
Tarski, A. (1955): "A Lattice-Theoretical Fixpoint Theorem and its Applications," *Pacific J. Math.*, *5*, 285–309.
Tarski, A. (1956): *Logic, Semantics, Metamathematics*, Papers from 1923 to 1938, Oxford University Press, N.Y.
Turing, A. M. (1936): "On Computable Numbers, With an Application to the Entscheidungsproblem," *Proc. London Math. Soc.* (2), *42*, 230–265; "A Correction", *ibid.*, *43* (1937), 544–546.
Turing, A. M. (1937): "Computability and λ-Definability," *J. Symb. Logic*, *2*, 153–163.
Yasuhara, A. (1971): *Recursive Function Theory and Logic*, Academic Press.

Notation Index

Author Index

Subject Index

Texts and Monographs in Computer Science

Hans W. Gschwind and Edward J. McCluskey
Design of Digital Computers
An Introduction
2nd Edition. 1975. viii, 548p. 375 illus. cloth

Brian Randell, Ed.
The Origins of Digital Computers
Selected Papers
2nd Edition. 1975. xvi, 464p. 120 illus. cloth

Jeffrey R. Sampson
Adaptive Information Processing
An Introductory Survey
1976. x, 214p. 83 illus. cloth

Arto Salomaa and Matti Soittola
Automata-Theoretic Aspects of Formal Power Series
1978. x, 171p. cloth

Suad Alagić and Michael A. Arbib
The Design of Well-Structured and Correct Programs
1978. x, 292p. 14 illus. cloth

Peter W. Frey, Ed.
Chess Skill in Man and Machine
1978. xi, 225p. 55 illus. cloth

David Gries, Ed.
Programming Methodology
A Collection of Articles by Members of IFIP WG2.3
1978. xiv, 437p. 68 illus. cloth

Michael A. Arbib, A.J. Kfoury, and Robert N. Moll
A Basis for Theoretical Computer Science
(The AKM Series in Theoretical Computer Science)
1981. vii, 220p. 49 illus. cloth